ASPECTS OF POLYURETHANES

Edited by **Faris Yilmaz**

Aspects of Polyurethanes
http://dx.doi.org/10.5772/65991
Edited by Faris Yilmaz

Contributors

Juan A. Galbis, M. De Gracia García-Martín, M. Violante De Paz, Elsa Galbis, Joanna Paciorek-Sadowska, Emriye Perrin Akcakoca Kumbasar, Cigdem Akduman, Michal Petrů, Ondřej Novák, Jitang Fan, Ayesha Kausar, Pierre Mertiny, Andre McDonald, Sayed Hossein Ashrafizadeh, Suman Thakur, Jin-Lian Hu

Published by InTech
Janeza Trdine 9, 51000 Rijeka, Croatia

Publishing Process Manager Romina Rovan
Technical Editor SPi Global
Cover InTech Design team

First published September, 2017
Printed in Croatia
Legal deposit, Croatia: National and University Library in Zagreb

Additional hard copies can be obtained from orders@intechopen.com

Aspects of Polyurethanes, Edited by Faris Yilmaz
p. cm.
Print ISBN 978-953-51-3545-6
Online ISBN 978-953-51-3546-3

For EU product safety concerns:
IN TECH d.o.o., Prolaz Marije Krucifikse Kozulić 3, 51000 Rijeka, Croatia,
info@intechopen.com or visit our website at intechopen.com.

Meet the editor

Dr. Faris Yilmaz was born in Palestine on October 07, 1969. After precollege education in Jenin (West Bank), he was accepted to Chemical Engineering Department of the Middle East Technical University (METU) in Ankara, Turkey. There he obtained his BSc degree in Chemical Engineering and MSc and PhD degrees in Polymer Science and Technology. His research has mainly concentrated on the conducting polymers and their solution properties, composites, and nanocomposites. He has publications in different polymer journals. He is married and has two children.

Contents

Preface

Polyurethanes, invented back in the 1930s by Professor Dr. Otto Bayer (1902–1982), are one of the most versatile plastic materials. There are various types of polyurethanes, which look and feel very different from each other. The nature of the chemistry allows them to be adapted to solve challenging problems, to be molded into unusual shapes, and to enhance industrial and consumer products by adding comfort, warmth, and convenience to our lives. They are used in a variety of products, from coatings and adhesives to shoe soles, mattresses, and foam insulation. Yet, the basic chemistry of each type is essentially the same.

Polyurethanes are formed by reacting a polyol (an alcohol with more than two reactive hydroxyl groups per molecule) with a diisocyanate or a polymeric isocyanate in the presence of suitable catalysts and additives. Because a variety of diisocyanates and a wide range of polyols can be used to produce polyurethane, a broad spectrum of materials can be produced to meet the needs of specific applications.

During World War II, a widespread use of polyurethanes was first seen, when they were used as a replacement for rubber, which at the time was expensive and hard to obtain. During the war, other applications were developed, largely involving coatings of different kinds, from airplane finishes to resistant clothing.

Subsequent decades saw many further developments, and today we are surrounded by polyurethane applications in every aspect of our everyday lives. While polyurethane is a product that most people are not overly familiar with, as it is generally "hidden" behind covers or surfaces made of other materials, it would be hard to imagine life without polyurethanes.

In the present book, we attempt to collect different topics and studies related to polyurethanes and their applications. In the first chapter, a discussion is done to illustrate a polyurethane-epoxy interpenetrating polymer network. Electrospun polyurethane nanofibers are illustrated in the second chapter. The third chapter is about polyurethane: a shape memory polymer (SMP). Also, measurement and numerical modeling of mechanical properties of polyurethane foams are discussed in the fourth chapter. Polyol containing boron atoms as a compound that reduces flammability of rigid polyurethane-polyisocyanurate foams; erosive and abrasive wear resistance of polyurethane liners; bio-based polyurethanes from carbohydrate monomers; and dynamic mechanical behavior of polymer materials are discussed in the fifth, sixth, seventh, and eighth chapters, respectively.

Dr. Faris Yilmaz
FNSS Savunma Sistemleri A.Ş.
Ankara, Turkey

Chapter 1

Polyurethane/Epoxy Interpenetrating Polymer Network

Ayesha Kausar

Additional information is available at the end of the chapter

http://dx.doi.org/10.5772/67678

Abstract

Polyurethane is a versatile thermoplastic polymer with a range of characteristics (tensile strength, chemical resistance, thermal stability, and processability) as coating, elastomer, foam, and fiber for technical application. Epoxy resin, on the other hand, is thermoset having fine mechanical, chemical, and adhesion properties to be utilized as adhesives, coatings, and matrix for advanced composite materials. However, epoxy resins are rigid and brittle in nature and have poor crack resistance. To overcome these problems, polyurethane phase has been introduced in the epoxy network for toughening. Considerable research has been carried out to introduce a second reactive polymer in the matrix to generate interpenetrating polymer network (IPN). Consequently, mechanical properties, glass transition behavior, thermal resistance, and damping features of polyurethane have been enhanced by introducing epoxy to form polyurethane/epoxy interpenetrating polymer network structure. Different modifiers have been employed to modify the properties of polyurethane/epoxy IPNs such as montmorillonite nanoclay, fibers, fly ash, conducting polymers, etc. The polyurethane/epoxy cross-linked networks have shown a range of high-performance application in ion-exchange resins, aircraft, engineering materials, biomedical devices, and other commercial IPN products.

Keywords: polyurethane, epoxy, network, IPN, application

1. Introduction

Polyurethane (PU) has been used in several technical applications due to high tensile strength, chemical resistance, processability, and mechanical properties [1, 2]. In polyurethane, hydrogen bonding is sufficient to produce a physical link between polymer chains to enhance overall improved properties. Phase separation in PU has been found to influence by hard and soft segment structures, molecular weight, polydispersity, and crosslinking ability. Epoxy polymers are tough and flexible with good corrosion and chemical resistance. The epoxy reactions are highly temperature dependent, epoxy oligomer/monomer-dependent, and energetic [3, 4]. Polymerizations can be typically carried out at or above ambient temperature (90–130°C). Solvent-free reactions are also used. Recently, there has been considerable interest in the development of mixtures with a second reactive polymer to generate interpenetrating polymer network (IPN) [5]. Interpenetrating polymer network is often considered as a novel type of polymer alloy, also known as polyblend. IPNs of polyurethane and epoxy have been reported by several investigators [6, 7]. In order to improve mechanical properties, thermal resistance, and damping properties of polyurethane, epoxy has been introduced in polyurethane systems to form polyurethane/epoxy IPN structure. Polyblends of linear polymers as well as grafted IPNs due to covalent bonding between the polymers have been established. Semi-IPN has also been developed from linear and crosslinked polymers [8]. Glass transition behavior and morphology studies have been used to explore the effective IPN. Generally, IPNs show excellent engineering properties due to synergetic effect induced by the compatibility of individual components in PU/epoxy IPNs [9]. Polyurethane/epoxy IPNs have been widely applied to foams, coatings, fibers, leather, and other applications [10]. To broaden the applications of PU/epoxy IPNs, these network structures have been modified using various filler structures. In this chapter, initially brief outlook on polyurethane and epoxy is illustrated. Afterward, the interpenetrating polymer network is discussed with special focus on polyurethane/epoxy IPNs. Applications of these networks produced by using the two unique polymers (polyurethane and epoxy) have also been conversed.

2. Prolog to polyurethane

Polyurethane (PU) is one of the most important classes of thermoplastic polymers having versatile structural relevance. Polyurethane elastomers are segmented copolymers consisting of hard and soft segment domains [11]. Soft domains are consequential of a macrodiol, while hard segments are derived from diisocyanate (**Figure 1**). When a short chain compound commonly referred as chain extender is used, polyurethanes are considered as segmented. In shape memory polyurethane (an important class of PU), hard segments in polyurethane are also known as fixed phase, while soft segments are termed as reversible phase. In general, hard and soft segments are incompatible with one another to create microphase separation. This segregation is principally responsible for excellent mechanical and other physical properties. Phase separation between the segments has been found to influence by hard segment structure, soft segment structure, molecular weight, weight fraction, polydispersity, and crosslinking. Polyurethane exists in several forms such as rubbery materials, liquid, soft solids, and thermoplastic. Few types of polyurethanes also exist as thermoset materials. A wide range of potential applications of polyurethanes have been achieved due to tailoring the essential features of PU. General applications of PU range from foam mattress to medical implant to engineering components [12, 13]. The advantage of the choice of polyurethane in advance applications is the ease of synthesis, processability, tailorability, chemical nature of hard and soft segments, and phase separation properties. Hydrogen bonding phenomenon

http://dx.doi.org/10.5772/67678

also occurs in PU chains, which offers physical crosslinking points between polymer chains. Condensation polymerization of *N*-(4-hydroxybenzal)*N'*(4-hydroxyphenyl)thiourea and methylene diisocyanate (MDI) has been studied [14]. The polyurethanes were used as an effective adsorbent for toxic metal ions. Solution precipitation route was used to prepare conducting polyurethane [15]. Several blends of polyurethanes have been studied. For example, polyurethane/polyaniline (PU/PANI) blend has been studied with enhanced electrical properties [16]. The effect of moisture on glass transition temperature (T_g) of polyurethane has also been studied [17].

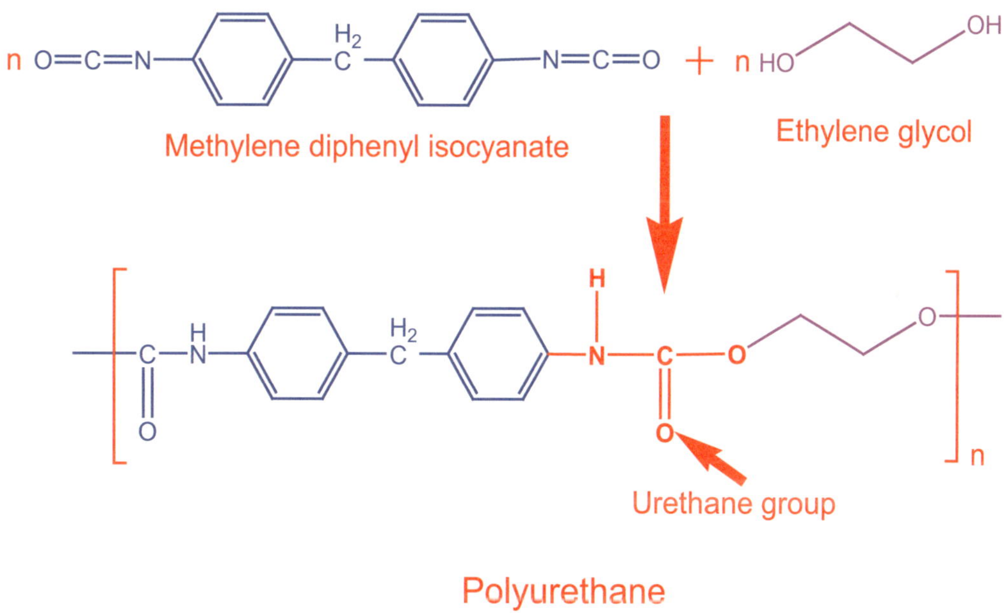

Figure 1. Chemical structure of PU elastomer.

3. Epoxy thermoset

Epoxy is generally illustrated by three-membered rings known as epoxy or oxirane or ethoxyline group. Epoxy resin is a prepolymer having more than one epoxide group. It is generally a low molecular weight compound. The most common type of epoxy resin is bisphenol-A epoxy resin (**Figure 2**). It is a type of epoxy resin produced by the reaction of epichlorohydrin with bisphenol-A in the presence of a basic catalyst. The properties of epoxy resin depend on the number of monomers in the epoxy chain [18]. Epoxy with low molecular weight usually has high viscosity and exists in the liquid state. High molecular weight epoxy occurs in solid state [19]. Depending upon the chemical structure of epoxy, they may have excellent electrical properties, thermal stability, UV stability, and weather ability [20]. In addition to linear epoxy resins, they can be cycloaliphatic, tri-functional, and tetra-functional epoxy resins [21, 22]. Various types of epoxy resins are shown in **Figure 3**.

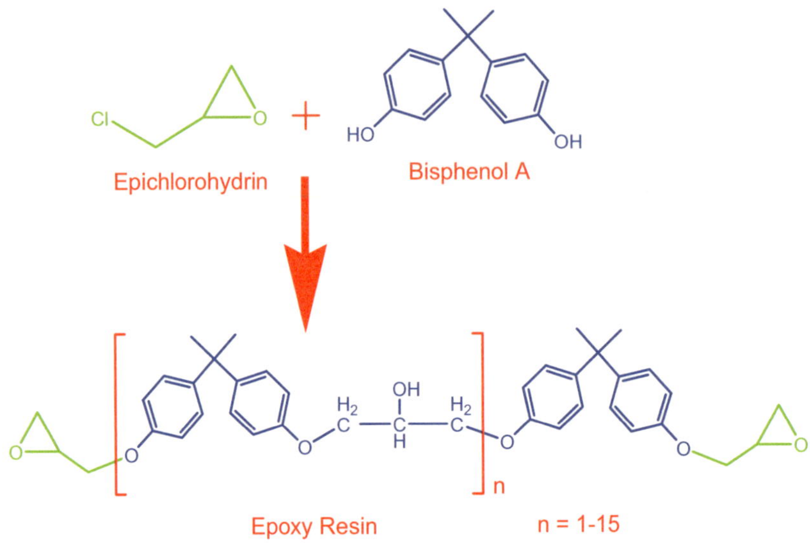

Figure 2. Preparation of bisphenol-A epoxy.

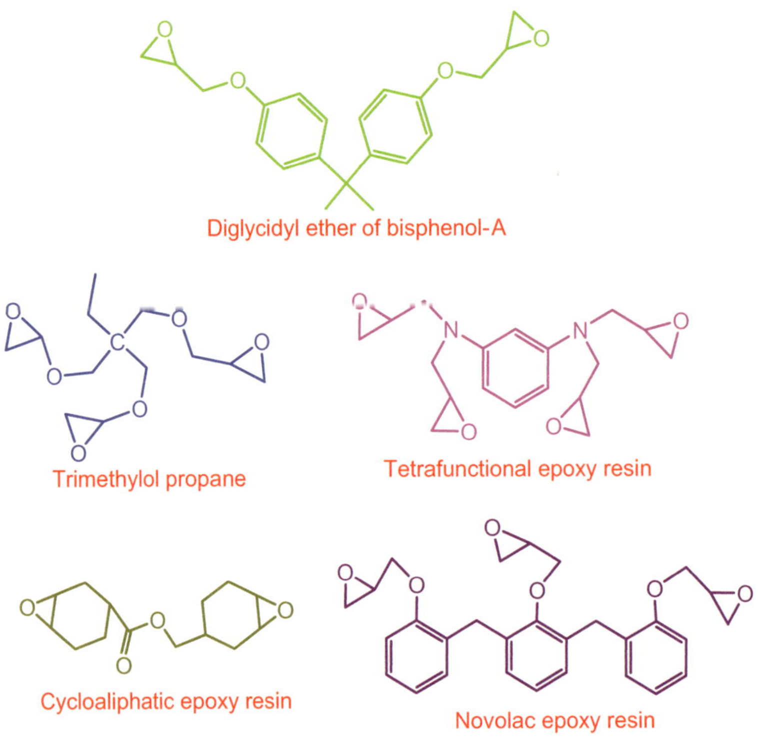

Figure 3. Different types of epoxy resin.

Among these types, tetra-functional epoxy resin has high crosslinking density and high thermal resistance. Novolac epoxy resin is produced by the reaction of aromatic novolac resin with epichlorohydrin. It has high crosslinking density due to the bulk of epoxide groups. Because of crosslinking property, it has excellent chemical, thermal, and solvent resistance properties [23]. Another other phenomenon is hardening of epoxy resins. The thermosetting resins are hardened using extensive range of hardening agents. The properties of thermosets also depend on the specific combination of hardening agents and epoxy resins constituting a system [24]. Molecular structure of hardening agents affects the final structure and properties particularly the glass transition temperature of epoxy. The hardening agents are also of various types such as amine, anhydride, and catalytic hardening agents [25]. Different types of amine hardening agents have been employed in epoxies [26]. The reactions between amine hardener and the epoxy are generally nucleophilic addition. These hardeners provide excellent chemical, physical, and electrical properties to the epoxy systems [27]. Hardening can be performed at room temperature, or in the presence of light, or heat. Room temperature hardeners provide high tensile properties, electrical conductivity, thermal resistance, and low T_g to epoxy systems [28]. Aliphatic polyamines, aromatic amines, and alicyclic polyamines are generally room temperature hardeners. Hardening time has been reduced using photocuring process, relative to heat curing or room temperature process [29].

4. Interpenetrating polymer network (IPN)

Interpenetrating polymer network (IPN) is a unique type of polyblend. IPNs can be simply defined as a mixture of two or more crosslinked polymeric networks [30]. The interpenetrating polymer network can be generated using physical or chemical interlocking between the polymer chains [31, 32]. IPNs can be considered as the crosslinking of one polymer component in the presence of another polymer component to form crosslinked polymer network. In this regard, fully formed interpenetrating polymer network and semi-interpenetrating polymer network have been identified [33]. Fully formed interpenetrating polymer network are acknowledged by the presence of crosslinks in both the network polymers, and the whole polymer components are crosslinked. Semi-IPN exists when one of the components is crosslinked and other is linear or non-crosslinked [34]. Semi-IPN is also referred to as pseudo-IPN. In *in situ* crosslinking, results in polymer chains are well interlocked. The full IPN and semi-IPN are shown in **Figure 4**. IPNs are sometimes confused with the polymer blends, block, graft, or crosslinked copolymers. However, there exist differences. An IPN can be distinguished from block copolymers as IPNs swell in the presence of solvents but does not dissolve. Moreover, IPNs exhibit characteristic morphologies. Polyurethane has been applied in elastomer, leather, foam, coatings, and fibers. Epoxy resin has been introduced in polyurethane systems to form epoxy/polyurethane interpenetrating polymer network structure. Various methods have been used to prepare and modify the properties of epoxy/polyurethane IPNs.

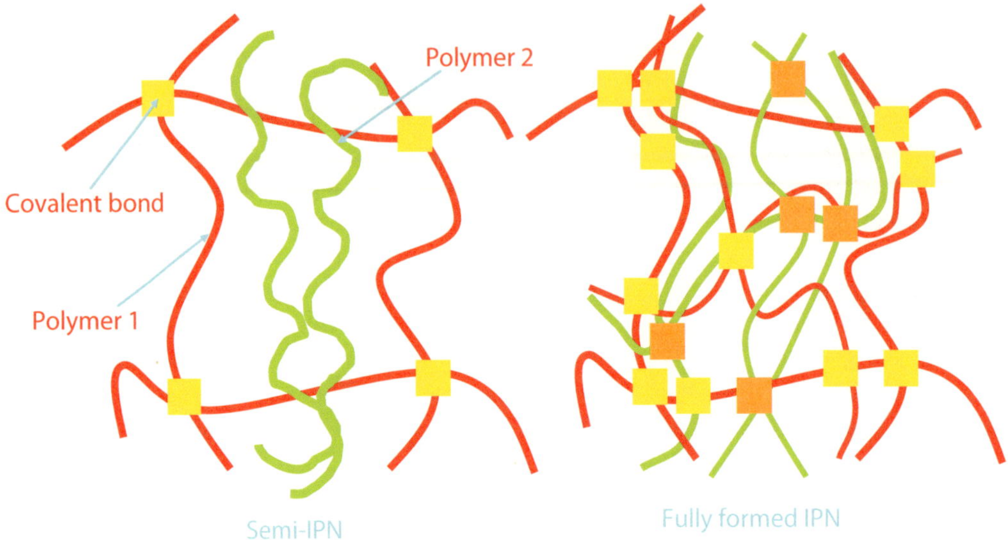

Figure 4. Interaction in IPNs.

5. Polyurethane/epoxy interpenetrating polymer network

Due to high modulus, strength, and mechanical properties, epoxy resins have been used in high-performance structural composites. However, engineering applications of epoxy resins have been limited in several cases such as damping materials [35]. In contrast, polyurethane is a flexible and elastic polymer with low mechanical strength [36]. PU prepolymers have been prepared and incorporated into epoxy resin to form IPNs. The mechanical properties of PU/epoxy IPNs largely depend on the amount of polyurethane in the blend network. Jin et al. [6] investigated tensile properties of soybean oil-based PU/epoxy IPNs. The tensile strength and tensile modulus of PU/EP IPNs with 5–20 mass% PU were lower than pure epoxy. This means that the addition of PU turned epoxy to the rubbery elastomer increased the elongation at break by 13-fold as compared to pure epoxy. However, increasing the amount of epoxy increased tensile strength and tensile modulus drastically. IPNs basically integrate the structure and properties of these versatile polymers (epoxy and polyurethane). Integrated performance of epoxy/polyurethane networks have been improved by the structural modification. The glass transition temperature of PU/epoxy IPNs provides important information about the miscibility of the blend components and blend compatibility. Moreover, the glass transition temperature of PU/epoxy IPNs also provides information about the cure rate of the reaction. At the beginning of cure reaction, T_g of PU/epoxy IPNs is usually lower than that of neat epoxy. The cure rate of epoxy is usually slower than that of PU. Some of the epoxide groups remain unreacted and act as plasticizer leading to lower T_g values of IPNs at the beginning of the cure reaction. When reaction proceeds and an IPN is formed, this may result in an increase in T_g of IPNs. The increase in T_g can be attributed to the miscibility/formation of graft structure through the reaction of hydroxyl groups of epoxy with isocyanate [37].

6. Modified polyurethane/epoxy IPNs

Polyurethane/epoxy IPNs have been prepared by adopting several modifications. The reinforced PU and diglycidyl ether of bisphenol A (DGEBA) IPN composites have been prepared with aramid fibers. The mechanical properties have been found to improve [38]. The PU/epoxy IPNs exhibited higher tensile and Izod impact strength. Montmorillonite-filled polyurethane/epoxy IPN nanocomposites were prepared, and the influence of hydrogen bonding on free volume and miscibility of clay was studied [39]. Polyethylene glycol-based polyurethane and epoxy IPN composite filled with fly ash have also been studied [40]. Semiconductive polyaniline and polyurethane/epoxy nanocomposite have been prepared for electromagnetic interference (EMI) shielding and charge dissipation applications. The damping properties of the modified PU/epoxy IPN composites have also been studied [41]. T_g, contact angle, interfacial, and mechanical properties have been investigated [42]. Recently, Kausar and Rahman Ur [43, 44] reported modified epoxy/polyurethane interpenetrating networks and their composites. Damping of vibration is a critical problem in the design of structural materials because excessive vibration may cause damage to the surroundings or the material components. To solve this problem, polymeric and composite IPNs with high damping properties around glass transition temperatures have been focused [45]. Chen et al. [46] prepared a series of castor-oil-based polyurethane/epoxy resin graft IPNs modified by hydroxy-terminated liquid nitrile rubber (HTLN). **Figure 5** and **Table 1** show the damping properties of HTLN-modified PU/epoxy IPN composites at 10 Hz. The glass transition temperature (corresponding to the peak of tan δ) was shifted to higher values with HTLN addition compared with that of pure IPNs. T_g of 5% loaded composite was increased to 71°C relative to neat IPNs (68.2°C). The 5% HTLN-modified PU/epoxy also showed good damping properties.

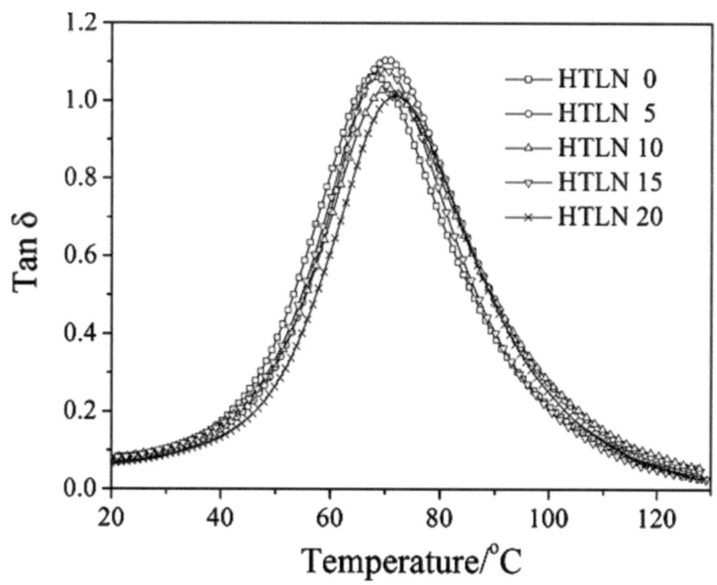

Figure 5. DMA curves of PU/EP IPN composites as a function of the HTLN content at 10 Hz [46].

The tensile strength of polyurethane/epoxy IPN composites filled with montmorillonite (MMT) has been studied [8]. According to **Figure 6**, 1% MMT resulted in the maximum tensile strength of the composites. The tensile strength of all the filled composites increased by about 40% relative to pure PU/epoxy IPNs. The results were attributed to the large interface area and strong interfacial adhesion between the IPN matrix and MMT. **Figure 7** shows the impact strength of MMT-filled PU/epoxy IPN composites. The impact strength primarily increased and then decreased with the increase in the MMT content. The impact strength reached higher values for 1 and 3% MMT contents. The results showed strong mutual interactions between the filler and matrix, and thus the impact strength of MMT-modified composites was improved.

PU/epoxy/HTLN (%)	tan δ	T_g (°C)
50/50/0	1.063	68.2
50/50/5	1.105	71.0
50/50/10	1.031	70.9
50/50/15	1.081	70.0
50/50/20	1.041	71.9

Table 1. DMA data of PU/EP IPN and HTLN-modified PU/EP IPN composites at 10 Hz [46].

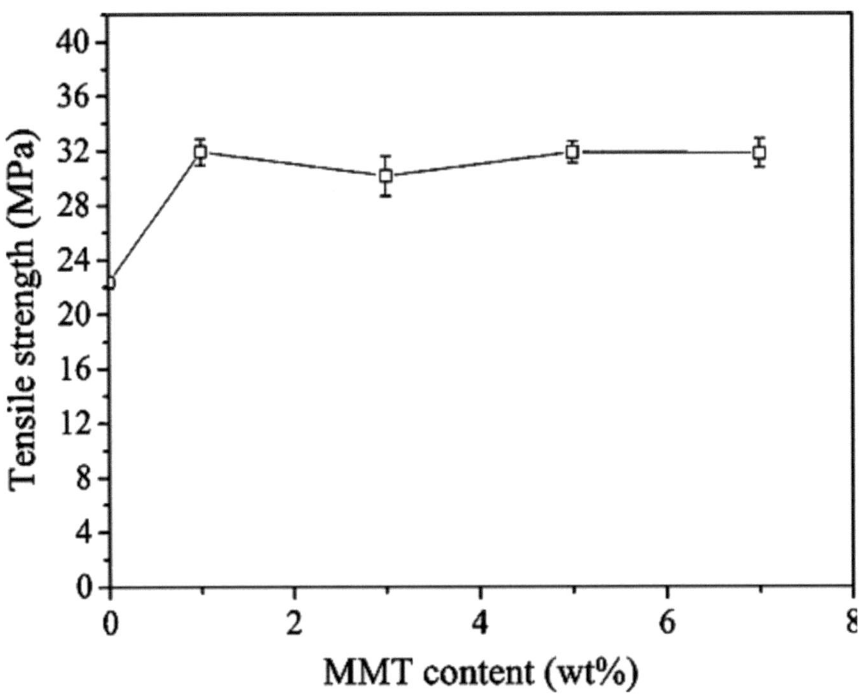

Figure 6. Tensile strength of MMT-filled PU/EP IPN composites as a function of the MMT content [8].

http://dx.doi.org/10.5772/67678

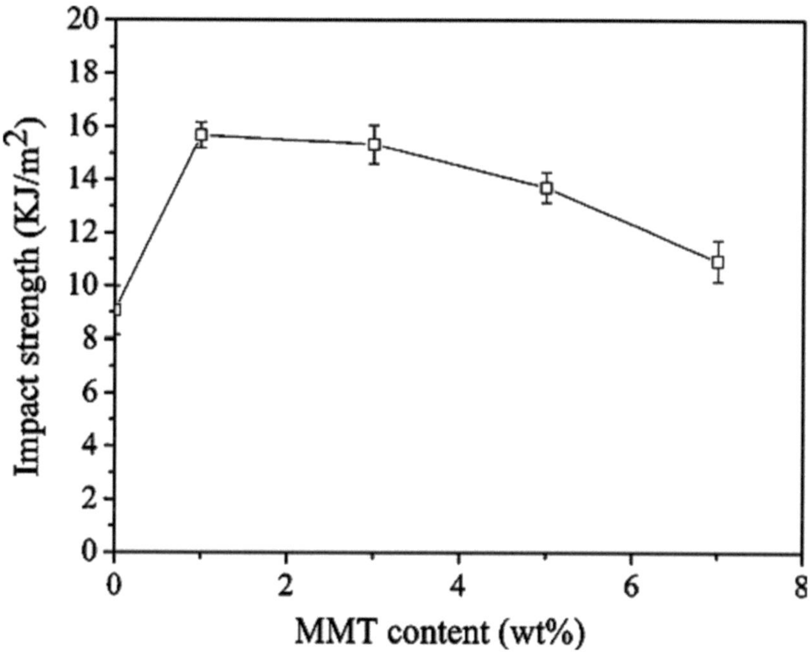

Figure 7. Impact strength of MMT-filled PU/EP IPN composites as a function of the MMT content [8].

7. Structure-property relationship in interpenetrating polymer network

The understanding of structure-property relationship in polyurethane/epoxy interpenetrating network is essential for a wide range of technical applications. Particularly, in the search of broadband damping materials, it is desirable to form IPNs with high loss region in shear or extension, covering the entire temperature and frequency range. Polyurethane/epoxy IPNs have been prepared for broadband damping materials. Epoxy and PU composites have been prepared to increase the crosslink density to improve polymer toughness. Epoxies have low toughness and poor crack resistance at room temperature due to high crosslinking density. Therefore, the literature focused on the matter of toughening of epoxies. Thermoplastic polymers/rubbers have been added to epoxy resins to improve the toughness of epoxy [47, 48]. Interpenetrating polymer network has been formed by the combination of crosslinked polymer networks in which at least one polymer is crosslinked in the immediate presence of the other. The mechanical properties of IPN structures are superior to the neat polymers. The presence of intermolecular hydrogen bonding between the hydroxyl group of epoxy and isocyanate group in PU plays an important role in interlocking network formation. IPNs of polyurethane and epoxy resin are very efficient for improving the fracture properties of the epoxy resin. Epoxy resin/polyurethane IPN nanocomposites with various contents of organophilic montmorillonite have been prepared through an *in situ* intercalation method [49]. The addition of PU to the epoxy matrix has remarkably increased the fracture toughness by 49% of pure epoxy, while the addition of clay improved toughness by 55% of pure epoxy. The resulting IPN shows excellent physical strength and low density. The low density may help in matching the specific acoustical impedance of the composite to the sea water [50]. The material can be acoustically transparent at the operating frequency range with high impact loading [51]. A successful operation of high-frequency sonar arrays needs an acoustic window with minimal interference, acoustical signals, and sufficient rigidity. To meet the demands of current designs of window materials, PU/epoxy composite materials are desired [52]. During interpenetrating polymer network (IPN) formation, qualitative control and relationship between temperature and polymerization kinetics are essential to understand and manage. Furthermore, the relationship of IPN formation with monomer type, concentration, temperature, and other reaction conditions should be considered for the optimum design of high-performance systems.

8. Application of IPNs

As described in the preceding sections, polyurethane/epoxy IPNs are of interest because they own enhanced glass transition temperature, impact strength, tensile strength, and damping properties relative to the neat/individual polymers [7, 8, 37, 46–49]. These properties are definitely superior to the polymer blends, which are usually obtained by blending the polymers together. If two homopolymers are blended, in most of the cases two distinct glass transition temperatures are observed. A major advantage is that the IPN may display a broad glass transition temperature. In this regard, Zhang and Hourston et al. [37] prepared a series of rigid interpenetrating polymer networks of rosin-based polyurethane and epoxy resin by a simultaneous polymerization technique. The chemical structure, dynamic mechanical properties, and morphology of the materials were investigated using relevant techniques. The PU/epoxy IPN showed a single broad glass transition over a wide range of composition, with the tan δ peaks of the IPN shifting to a lower temperature. This implies that the PU/epoxy IPN foam system was miscible over a wide range of composition. In other words, the single broad glass transition temperature indicated good compatibility of the two polymers contributing to IPNs. As the epoxy graft content was increased, tan δ peaks move toward the peak of neat epoxy. This broad glass transition temperature is highly advantageous for energy absorption and vibration damping [53, 54]. In drug delivery, IPNs have been used to maximize the therapeutic benefits of the drug. Moreover, biologically active materials have been prepared when controlled release is desirable. The physiochemical properties such as drug diffusivity, erosion rate, and controlled dissolution can be tailored *in vivo* through selection of the materials, composition and crosslink density. Another important application of IPNs is in dental applications. IPNs have been used as a synthetic teeth and cavity filler. Moreover, the IPN offers a number of advantages such as less temperature sensitivity and stronger bonding to the tooth. In engineering applications, the IPN has advantage over homopolymers or homopolymer composites. Using IPNs, the material properties can be tailored to a higher degree at different stages of polymerization. However, IPN complexity may arise from processing conditions affecting the material properties, kinetics, and thermodynamic instabilities driving phase separation. The use of these IPNs has also been exploited in hydrodynamic machines such as water turbine pumps.

http://dx.doi.org/10.5772/67678

PU/epoxy IPNs have the ability to prevent the damage of cavitation corrosion. These IPNs are used to form cavitation corrosion resistant coating due to good adhesion to metals, abrasive resistance, water resistance, elasticity, toughness, and damping property. PU/epoxy IPNs have also been used in several applications such as thermally conductive adhesives in electronic components. The thermal conductivity of composites can be improved by using polyurethane/epoxy fully formed IPNs [55, 56]. To explore and implement more technical applications of PU/epoxy IPNs, fabrication processes, processing conditions, and developing trends must be focused in future.

9. Conclusions

The principal routes for the formation of IPNs are successive and simultaneous polymerization of the two polymer components. Polyurethane elastomers are segmented copolymers consisting of soft segment domains derived from a macrodiol and hard segment domains derived from a diisocyanate and chain extender. Usually, the two segments are incompatible, resulting in microphase separation, which is responsible for fine mechanical properties. Epoxy resins are well known for unique properties such as high mechanical strength, chemical resistance, and outstanding surface adhesion. However, they are rigid and brittle in nature, and have poor crack resistance, which prevent its engineering applications. To overcome these problems, toughening of epoxy resins with polyurethanes has been performed. Various factors affect the final properties of IPNs such as hard and soft segment structures, molecular weight, polydispersity, crosslinking, and phase separation. Consequently, the design of hard and soft segment structures affects the structure of PU prepolymer. The molecular weight of PU prepolymer also influences the formation of IPN structure. Phase separation is truly dependent on the content of polyurethane and epoxy contributing to the IPN network. Furthermore, the degree of crosslinking defines the glass transition temperature of the final PU/epoxy IPN. Two types of forces may form the IPN network in polymers, i.e., primary (chemical bond) and secondary (vander Waals). The presence of intermolecular hydrogen bonding between the hydroxyl group in epoxy and the isocyanate group in PU plays an important role in increasing network interlocking in IPN formation. Depending on the specific components selected for IPN formation, surface free energy of the blend system, and other structural parameters, there are several technical applications identified for epoxy/polyurethane systems.

Author details

Ayesha Kausar[1,2*]

*Address all correspondence to: asheesgreat@yahoo.com

1 Nanosciences Division, National Center For Physics, Quaid-i-Azam University, Islamabad, Pakistan

2 Department of Chemistry, Quaid-i-Azam University, Islamabad, Pakistan

References

[1] Usman A, Zia KM, Zuber M, Tabasum S, Rehman S, Zia F. Chitin and chitosan based polyurethanes: A review of recent advances and prospective biomedical applications. International Journal of Biological Macromolecules. 2016; **86**: 630-645. DOI: 10.1016/j.ijbiomac.2016.02.004.

[2] Mahmood N, Yuan Z, Schmidt J, Xu CC. Depolymerization of lignins and their applications for the preparation of polyols and rigid polyurethane foams: A review. Renewable and Sustainable Energy Reviews. 2016; **60**: 317-329.DOI: 10.1016/j.rser.2016.01.037

[3] Kausar A. Role of thermosetting polymer in structural composite. American Journal of Polymer Science & Engineering. 2017; **5**: 1-12.

[4] Qu T, Yang N, Hou J, Li G, Yao Y, Zhang Q, He L, Wu D, Qu X. Flame retarding epoxy composites with poly (phosphazene-co-bisphenol A)-coated boron nitride to improve thermal conductivity and thermal stability. RSC Advances. 2017; **7**: 6140-6151. DOI: 10.1039/C6RA27062J.

[5] Chen S, Wang Q, Wang T, Pei X. Preparation, damping and thermal properties of potassium titanate whiskers filled castor oil-based polyurethane/epoxy interpenetrating polymer network composites. Materials & Design. 2011; **32**: 803-807. DOI: 10.1016/j.matdes.2010.07.021.

[6] Jin H, Zhang Y, Wang C, Sun Y, Yuan Z, Pan Y, Xie H, Cheng R. Thermal, mechanical, and morphological properties of soybean oil-based polyurethane/epoxy resin interpenetrating polymer networks (IPNs). Journal of Thermal Analysis and Calorimetry. 2014; **117**: 773-781. DOI: 10.1007/s10973-014-3849-5.

[7] Wang C, Jia J. Damping and mechanical properties of polyol cross-linked polyurethane/epoxy interpenetrating polymer networks. High Performance Polymers. 2014; **26**: 240-244. DOI: 10.1177/0954008313508421.

[8] Chen S, Wang Q, Wang T. Damping, thermal, and mechanical properties of montmorillonite modified castor oil-based polyurethane/epoxy graft IPN composites. Materials Chemistry and Physics. 2011; **130**: 680-684. DOI: 10.1016/j.matchemphys.2011.07.044.

[9] Simendić JB, Bjelović Z, Jovanović SS, Aleksić V, Valentova H, Szecseny KM, Krakovsky I. Thermal stability and damping properties of polyurethane hybrid material based on castor oil. Contemporary Materials. 2014; **1**: 64-68. DOI: 10.7251/cm.v1i5.1500.

[10] Prashantha K, Rashmi BJ, Pai KV. Preparation and characterization of Poly (2-hydroxyethyl methacrylate) and castor oil based uralkyds interpenetrating polymer networks. Polymers & Polymer Composites. 2015; **23**: 223.

[11] Oprea S. Synthesis and properties of polyurethane elastomers with castor oil as crosslinker. Journal of the American Oil Chemists Society. 2010; **87**: 313-320. DOI: 10.1007/s11746-009-1501-5.

[12] Ahmed N, Kausar A, Muhammad, B. Advances in shape memory polyurethanes and composites: A review. Polymer-Plastics Technology and Engineering. 2015; **54**: 1410-1423. DOI: 10.1080/03602559.2015.1021490.

[13] Kausar A. Review on technological significance of photoactive, electroactive, pH sensitive, water-active, and thermo-responsive polyurethane materials. Polymer-Plastics Technology and Engineering. 2017; **56**: 606-616. DOI: 10.1080/03602559.2016.1233279.

[14] Kausar A, Zulfiqar S, Sarwar MI. High performance segmented polyurethanes derived from a new aromatic diisocyanate and polyol. Polymer degradation and stability. 2013; **98**: 368-376. DOI: 10.1016/j.polymdegradstab.2012.09.004.

[15] Sattar R, Kausar A, Siddiq M. Advances in thermoplastic polyurethane composites reinforced with carbon nanotubes and carbon nanofibers: A review. Journal of Plastic Film & Sheeting. 2015; **31**: 186-224. DOI: 10.1177/8756087914535126.

[16] Sattar R, Kausar A, Siddiq M. Thermal, mechanical and electrical studies of novel shape memory polyurethane/polyaniline blends. Chinese Journal of Polymer Science. 2015; **33**: 1313-1324. DOI: 10.1007/s10118-015-1680-5.

[17] Kausar A. Poly (urethane urea)/polythiophene/carbon black composite: Morphology, mechanical, and conducting shape memory behavior. Journal of Thermoplastic Composite Materials. 2016; DOI: 10.1177/0892705716679477.

[18] Brady RF. editors. Comprehensive Desk Reference of Polymer Characterization and Analysis. American Chemical Society: Oxford, New York. 2004, **81**: 1425, DOI: 10.1021/ed081p1425.2.

[19] Jin FL, Ma CJ, Park SJ. Thermal and mechanical interfacial properties of epoxy composites based on functionalized carbon nanotubes. Materials Science and Engineering. 2011; **528**: 8517-8522. DOI: 10.1016/j.msea.2011.08.054.

[20] Yoo MJ, Kim SH, Park SD, Lee WS, Sun JW, Choi JH, Nahm S. Investigation of curing kinetics of various cycloaliphatic epoxy resins using dynamic thermal analysis. European Polymer Journal. 2010; **46**: 1158-1162. DOI: 10.1016/j.eurpolymj.2010.02.001.

[21] Kwak GH, Park SJ, Lee JR. Thermal stability and mechanical behavior of cycloaliphatic–DGEBA epoxy blend system initiated by cationic latent catalyst. Journal of Applied Polymer Science. 2000; **78**: 290-297. DOI: 10.1002/1097-4628(20001010)78:2<290::AID-APP80>3.0.CO;2-9.

[22] Park SJ, Jin FL, Lee JR. Thermal and mechanical properties of tetrafunctional epoxy resin toughened with epoxidized soybean oil. Materials Science and Engineering. 2004; **374**: 109-114. DOI: 10.1016/j.msea.2004.01.002.

[23] Lee MC, Ho TH, Wang CS. Synthesis of tetrafunctional epoxy resins and their modification with polydimethylsiloxane for electronic application. Journal of Applied Polymer Science. 1996; **62**: 217-225. DOI: 10.1002/(SICI)1097-4628(19961003)62:1<217::AID-APP25>3.0.CO;2-0.

[24] Bascom W, Cottington, R, Jones P, Peyser R. The fracture of epoxy- and elastomer-modified epoxy polymers in bulk and as adhesives. Journal of Applied Polymer Science. 1975; **19**: 12545-2562. DOI: 10.1002/app.1975.070190917.

[25] Hsu YG, Lin KH, Lin TY, Fang YL, Chen SC, Sung YC. Properties of epoxy-amine networks containing nanostructured ether-crosslinked domains. Matererials Chemistry and Physics. 2012; **132**: 688-702. DOI: 10.1080/1536383X.2016.1172067.

[26] Ferdosian F, Ebrahimi M, Jannesari A. Curing kinetics of solid epoxy/DDM/nanoclay: Isoconversional models versus fitting model. Thermochim Acta. 2013; **568**: 67-73. DOI: 10.1016/j.tca.2013.06.001.

[27] Fan M, Liu J, Li X, Cheng J, Zhang J. Curing behaviors and properties of an extrinsic toughened epoxy/anhydride system and an intrinsic toughened epoxy/anhydride system. Thermochim Acta. 2013; **554**: 39-47. DOI: 10.1016/j.tca.2012.12.007.

[28] Ahmad Z, Ansell MP, Smedley D. Effect of nano-and micro-particle additions on moisture absorption in thixotropic room temperature cure epoxy-based adhesives for bonded-in timber connections. International Journal of Adhesion and Adhesives. 2010; **30**: 448-455. DOI: 10.1016/j.ijadhadh.2010.04.001.

[29] Jiang W, Jin FL, Park SJ. Thermo-mechanical behaviors of epoxy resins reinforced with nano-Al_2O_3 particles. Journal of Industrial and Engineering Chemistry. 2012; **18**: 594-596. Doi: 10.1016/j.jiec.2011.11.140.

[30] Ma C, Goang DY, Han JL, Hsieh KH. Interpenetrating polymer networks of polyurethane and furfuryl alcohol, IIa. Processability and properties of pultruded fiber-reinforced composites. Die Angewandte Makromolekulare Chemie. 1994; **218**: 41-52. DOI: 10.1002/apmc.1994.052180104.

[31] Kausar A. Estimation of thermo-mechanical and fire resistance profile of epoxy coated polyurethane/fullerene composite films. Fullerenes, Nanotubes and Carbon Nanostructures. 2016; **24**: 391-399. DOI: 10.1080/1536383X.2016.1172067.

[32] Kausar A. An investigation on epoxy/poly (urethane-amide)-based interpenetrating polymer network reinforced with an organic nanoparticle. Materials Research Innovations. 2016; DOI: 10.1080/14328917.2016.1265232.

[33] Tan F, Qiao X, Chen J, Wang H. Effects of coupling agents on the properties of epoxy-based electrically conductive adhesives. International journal of adhesion and adhesives. 2006; **26**: 406-413. DOI: 10.1016/j.ijadhadh.2005.06.005.

[34] Kim SC, Klempner D, Frisch KC, Frisch HL. Polyurethane interpenetrating polymer networks. II. Density and glass transition behavior of polyurethane-poly (methyl methacrylate) and polyurethane-polystyrene IPN's. Macromolecules. 1976; **9**: 263-266. DOI: 10.1021/ma60050a017.

[35] Jeevananda T. Synthesis and characterization of polyaniline filled PU/PMMA interpenetrating polymer networks. European Polymer Journal. 2003; **39**: 569-578. DOI: 10.1016/S0014-3057(02)00272-0.

[36] Lin SP, Han JL, Yeh JT, Chang FC, Hsieh KH. Composites of UHMWPE fiber reinforced PU/epoxy grafted interpenetrating polymer networks. European Polymer Journal. 2007; **4**: 996-1008. DOI: 10.1016/j.eurpolymj.2006.12.001.

[37] Zhang Y, Hourston DJ. Rigid interpenetrating polymer network foams prepared from a rosin-based polyurethane and an epoxy resin. Journal of Applied Polymer Science. 1998, **69**: 271-381. DOI: 10.1002/(SICI)1097-4628(19980711)69:2^#x003C;271::AID-APP8>3.0.CO;2-K.

[38] Han JL, Lin SP, Ji SB, Hsieh KH. Graft interpenetrating polymer networks of polyurethane and epoxy containing rigid rods in side chain. Journal of Applied Polymer Science. 2007; **106**: 3298-3307. DOI: 10.1002/app.26874.

[39] Zhu YC, Wang B, Gong W, Kong LM, Jia QM. Investigation of the hydrogenbonding structure and miscibility for PU/EP IPN nanocomposites by PALS. Macromolecules. 2006; **39**: 9441-9445. DOI: 10.1021/ma0621066.

[40] Wu GH, Gu J, Zhao X. Preparation and dynamic mechanical properties of polyurethane-modified epoxy composites filled with functionalized fly ash particulates. Journal of Applied Polymer Science. 2007; **105**: 1118-1126. DOI: 10.1002/app.26146.

[41] Chiou WC, Han JL, Lee SN. Synthesis and studies of the physical properties of polyaniline and polyurethane-modified epoxy composites. Polymer Engineering and Science. 2008; **48**: 345-354. DOI: 10.1002/pen.20944.

[42] Park SJ, Jin JS. Energetic studies on epoxy–polyurethane interpenetrating polymer networks. Journal of Applied Polymer Science. 2001; **82**: 775-780. DOI: 10.1002/app.1903.

[43] Kausar A. Nanodiamond tethered epoxy/polyurethane interpenetrating network nanocomposite: Physical properties and thermoresponsive shape-memory behavior. International Journal of Polymer Analysis and Characterization. 2016; **21**: 348-358. DOI: 10.1080/1023666X.2016.1156911.

[44] Kausar A, Rahman Ur A. Effect of graphene nanoplatelet addition on properties of thermo-responsive shape memory polyurethane-basednanocomposite. Fullerenes, Nanotubes and Carbon Nanostructures, 2016; **24**: 235-242. DOI: 10.1080/1536383X.2016.1144592.

[45] Buravalla VR, Remillat C, Rongong JA, Tomlinson GR. Advances in damping materials and technology. Smart Materials Bulletin. 2001, **8**: 10-13. DOI: 10.1016/S1471-3918(01)80184-2.

[46] Chen S, Wang Q, Wang T. Hydroxy-terminated liquid nitrile rubber modified castor oil based polyurethane/epoxy IPN composites: Damping, thermal and mechanical properties. Polymer Testing. 2011, **30**: 726-731. DOI: 10.1016/j.polymertesting.2011.06.011.

[47] Russell B, Chartoff R. The influence of cure conditions on the morphology and phase distribution in a rubber-modified epoxy resin using scanning electron microscopy and atomic force microscopy. Polymer. 2005, **46**: 785-798. DOI: 10.1016/j.polymer.2004.11.090.

[48] Liu Y, Zhang W, Zhou H. Mechanical properties of epoxy resin/hydroxyl-terminated polyester blends: Effect of two-phase structure. Polymer International. 2005. **54**: 1408-1415. DOI: 10.1002/pi.1862.

[49] Li J. High performance epoxy resin nanocomposites containing both organic montmorillonite and castor oil-polyurethane. Polymer Bulletin. 2006, **56**: 377-384. DOI: 10.1007/s00289-005-0492-0.

[50] Coughlin CS, Samuels MQ, Capps RN. Structure–property relationships in polymer interpenetrating networks for use as broad band damping materials. The Journal of the Acoustical Society of America. 1993; **94**: 1840-1840. DOI: 10.1121/1.407750.

[51] Thompson CM, Ting RY, Leimer LM. New fiber-reinforced polyurethane composites as a soft sonar window. The Journal of the Acoustical Society of America. 1993; **94**: 1840-1840. DOI: 10.1121/1.407752.

[52] Thompson CM, Ting RY, Jenkins LL, Chiou CC. Development of a new rigid composite for high-frequency sonar window applications. Acoustical Society of America Journal. 1993; **94**: 1840. DOI: 10.1121/1.407753.

[53] Cerqueira DA, Rodrigues Filho G, Assunção RM. A new value for the heat of fusion of a perfect crystal of cellulose acetate. Polymer Bulletin. 2006; **56**: 475-484. DOI: 10.1007/s00289-006-0511-9

[54] Yang J, Winnik, MA, Ylitalo D, Devoe RJ. Polyurethane-polyacrylate interpenetrating networks. 1. Preparation and Morphology Macromolecules. 1996; **29**: 7047-7054. DOI: 10.1021/ma9601373.

[55] Li Y, Mao S. Study on the properties and application of epoxy resin/polyurethane semi-interpenetrating polymer networks. Journal of Applied Polymer Science. 1996; **61**: 2059-2063. DOI: 10.1002/(SICI)1097-4628(19960919)61:12<2059::AID-APP2>3.0.CO;2-9.

[56] Zhanzheng B, Xiuli Z, Xuefang L, Shikai LU. Preparation and characterization of epoxy encapsulating materials toughened by polyurethane. Chemical Industry and Engineering Progress. 2009; **28**: 1010-1085.

Chapter 2

Electrospun Polyurethane Nanofibers

Cigdem Akduman and
Emriye Perrin Akçakoca Kumbasar

Additional information is available at the end of the chapter

http://dx.doi.org/10.5772/intechopen.69937

Abstract

The electrospinning process is highlighted with the ability of fabricating fibers with diameters on the nanometer scale, small inter-fibrous pore size and high porosity, vast possibilities for functionalization with high surface area to volume or mass ratio, ease of use and instrument setup, and adaptability. It attracted a great deal of attention due to its unique properties. More than 100 different polymers have been successfully electrospun into ultrafine fibers using this technique including synthetic polymers such as polyurethane (PU). Electrospun PU nanofiber mats exhibiting good mechanical properties may have a wide variety of potential applications in high-performance air filters, protective textiles, wound dressing materials, sensors, drug delivery, etc. This chapter deals with the electrospinning of polyurethane nanofibers and their potential applications.

Keywords: electrospinning, polyurethane, nanofibers, applications

1. Introduction

Electrospinning is a fiber spinning technique that is able to produce continuous ultrafine fibers from sub-micrometer to nanometer diameters, which uses electrostatic forces. The original ideas of using electrical potentials on the surfaces of droplets, amount of charge required for the deformation of droplets, and also apparatus for spraying of liquids by use of electrical charges can be traced back more than 100 years [1–3]. However, Formhals was recognized as the father of the electrospinning and he was the first to describe the operation of electrospinning in 1934 for producing polymer filaments by electrostatic repulsions between surface charges [4].

Despite these early studies, researches in nanofibers and electrospinning have received a great deal of attention, after 1990s, especially after Doshi and Reneker [5], Srinivasan and

Reneker [6], and Reneker and Chun [7], spun various kinds of polymers including polyethylene oxide, nylon, polyimide, DNA, polyaramid, and polyaniline, and were able to characterize their properties. Afterward this old technology rediscovered, refined, and expanded into non-textile applications. Electrospinning process is unique among other nanofiber fabrication techniques in terms of ease of use, vast possibilities of material selection and combination, and it has a potential for scale-up. This has led to electrospinning being considered as a key platform technology that will be studied to develop products for a wide range including drug delivery, tissue scaffolding, wound dressing, electronics, chemical sensors, filtration, and so on [8]. This rediscovery and attention are partly the results of leading-edge technology, especially, scanning probe microscopy and high-resolution electron microscopy, which enable the exploration of the "nanodimension."

2. Electrospinning

The electrospinning process simply constituted of a high voltage power supply, a spinneret and a deposition area named as collector generally covered with an aluminum foil. The potential difference between the spinneret and the collector leads to the stretch of the polymeric solution and creates a thin nanofiber jet from solution toward to the deposition area. During this electrospinning process, the solvent evaporates and ultrafine fibers are collected [9, 10]. Main process equipments and setup are presented in **Figure 1**.

The electrospinning process has attracted a great deal of attention due to the ability to fabricate fibers with diameters on the nanometer scale [11, 12], vast possibilities for surface functionalization [13, 14] with high surface area to volume or mass ratio, small inter-fibrous pore size and high porosity [11, 15, 16]. Technically, almost any soluble polymer with a sufficiently high molecular weight can be electrospun, and [17] various polymers have been successfully electrospun into nanofibers cost-effectively compared to the other methods.

The method can be applied to synthetic and natural polymers, polymer blends, and polymers loaded with chromophores, nanoparticles, or active agents, as well as to metals and ceramics [18]. A large number of inorganic salts, inorganic and organic particles, and carbon nanotubes (CNTs), can also be immobilized in polymer fibers [19]. More than 100 different polymers have been successfully electrospun into ultrafine fibers using this technique [10] including synthetic polymers such as PU [20–29].

3. Polyurethanes

PUs contains urethane group (–NH–(C=O)–O–) in common (**Figure 2a**) [30, 31], while most PUs are thermosetting polymers, in contrast to thermosetting polymers, thermoplastic polyurethanes (TPUs) melt when they are heated (**Figure 2b**) and are easy to use in manufacturing processes. Varying the structure of PUs, their properties can be varied in a wide range [32]. PUs are formed by reaction of polyisocyanates with hydroxyl-containing compounds.

http://dx.doi.org/10.5772/intechopen.69937

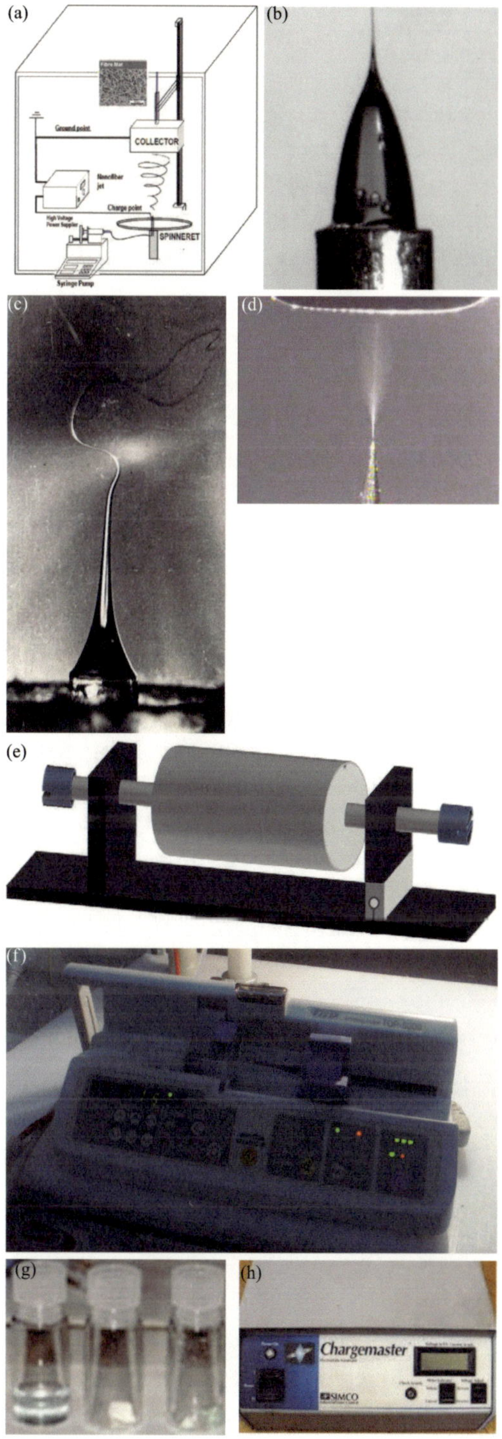

Figure 1. Electrospinning setup (a), Taylor cone (b), whipping (c), electrospining jet (d), collectors (e), single syringe pump (f), polymer solutions (g), and high voltage supply (h).

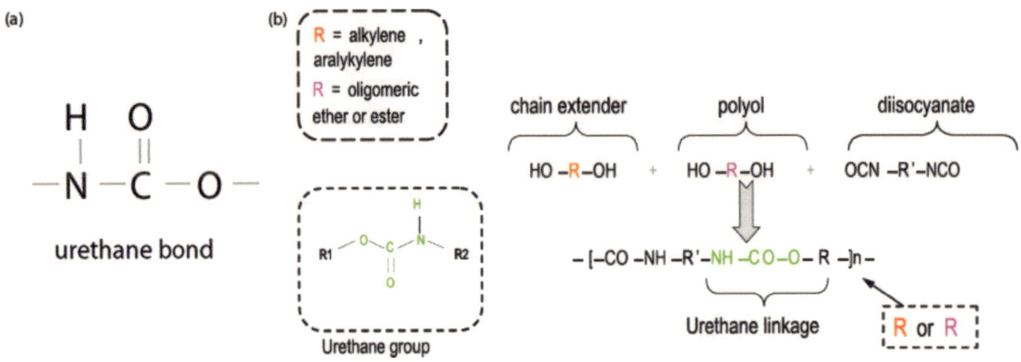

Figure 2. (a) Urethane group, (b) thermoplastic polyurethane general formula [35].

Desired properties can be tailored by selecting the type of isocyanate and polyols, or combination of isocyanates and combination of polyols [32, 33]. Strong intermolecular bonds make polyurethanes useful for diverse applications in adhesives and coatings, also in elastomers, foams, and medical applications because of their good biocompatibility [34].

The factors determining properties of a polyurethane elastomer are: structure of the polyol, type of diisocyanate, type of the chain extender, molar ratio NCO/OH, soft-segment concentration, molecular weight of the polyol and filler. In polyurethane elastomers, chains are linear, and cross-linking was achieved by physical bonds and hard domain formation. They flow when they are melted and harden by cooling (thermoplastic behavior). Displaying reversible cross-linking, domains are destroyed above the melting point of the hard phase, but are reformed when they get cooled. These materials are called "thermoplastic urethanes" (TPUs) [32]. In addition to the linear TPUs, obtained from difunctional monomers, branched or cross-linked thermoset polymers are made with higher functional monomers. Linear polymers have good impact strength, good physical properties, and excellent processibility, but limited thermal stability. On the other hand, thermoset polymers have higher thermal stability, but sometimes lower impact strength [33].

The vast selection of polyols, isocyanates, and chain extenders allows PUs to be varied from soft thermoplastic elastomers to adhesives, coatings, flexible foams, and rigid thermosets [25]. TPU elastomers are segmented block copolymers, comprising of hard- and soft-segment blocks. The soft-segment blocks are formed from long-chain polyester or polyether polyols and 4,4′,-methylenebis(phenyl isocyanate) (MDI); the hard segments are formed from short-chain diols, mainly 1,4-butanediol and MDI [33]. The unique properties of linear TPUs are attributed to their long-chain structure. TPUs are resilient elastomers of significant industrial importance, which possess a range of desirable properties such as elastomeric, resistant to abrasion, and excellent hydrolytic stability [36, 37].

PU is often chosen as a material for composing a nanoweb due to its chemical stability, mass transport, good mechanical properties, and also excellent nanofiber forming characteristic [24, 38]. Electrospun PU nanofiber mats exhibiting good mechanical properties may have a wide variety of potential applications in high-performance air filters, protective textiles, wound dressing materials, sensors, in biomedical applications, drug delivery, etc. [21, 37–44]. It is also frequently used in wound dressing studies because of its good barrier properties and oxygen permeability. It has reported that semipermeable dressings, many of which are PU, enhance wound healing [21, 39]. Degradable and biocompatible aliphatic PU could also be formed into scaffolds via melt electrospinning [42]. Beside these biomedical applications, the PU nanofiber filters were prepared by electrospinning process based on 3D particle filtration modeling and some theoretical predictions were obtained for the filtration efficiency [45]. In another study, PU cationomers (PUCs) containing different amounts of quaternary ammonium groups were synthesized and electrospun into non-woven nanofiber mats for use in antimicrobial nanofilter applications [46].

Varying the structure of PUs, the properties of PUs can vary in a wide range [32]. The flexibility to tailor the structure during processing is one of the main advantages of PUs over other types of polymers. A lot of different types of PU used in electrospinning process, some of them were synthesized before electrospinning according to the researchers intended use and some of them were used as they received.

This chapter deals with the electrospinning of PU nanofibers including TPUs and their potential applications. The nanofiber morphology and influence of experimental parameters including the solution concentration, flow rate, collection distance, and electric voltage are discussed in terms of electrospinning of PU nanofibers and potential applications of PU nanofibers are reviewed.

4. Electrospinning of PU nanofibers

There are many parameters that affect the electrospinning process and the resultant fiber morphology. Several authors investigated the effects of solution and process parameters such as material composition, concentration, rheological properties, applied voltage, tip-to collector distance, collector types on the resultant PU nanofibers [26, 27, 47–49]. Among these parameters, polymer solution properties have the most significant influences on the process and the resultant fiber morphology [9], since viscoelastic and gravitational forces play a major role. Viscoelastic force depends on polymer solution concentration, average molecular weight of the polymer, final viscosity, and surface tension of the solution [47].

4.1. Solvents for electrospun PU nanofibers

Electrospinning solution properties are also directly related to the solvent type. The type of solvent and their concentration influenced the morphology of electrospun nanofiber mat significantly. Process involves the stretching of the solution caused by repulsion of the charges at its surface. Thus, if the conductivity of the solution is increased, more charges can be carried by the electrospinning jet [13], which increases the stretching of the polymer solution. As much as solution conductivity, the dielectric constant of a solvent has also a significant influence in electrospinning process. Higher dielectric property reduced the bead formation and the diameter of the resultant electrospun fibers [47].

Most commonly used solvents that dissolve PUs are highly polar organic solvents such as N,N-dimethylformamide (DMF), dimethylsulfoxide (DMSO), N-Methylpyrrolidone and tetrahydrofuran (THF) [50, 51]. Ketones such as acetone, methyl ethyl ketone, cyclohexanone are partial solvents for TPUs. Aliphatic alcohols such as ethanol and isopropanol cause a slight swelling; aliphatic esters such as ethyl acetate and butyl acetate strongly cause severe swelling of TPUs [50]. Investigation of the effects of solvents in electrospinning of TPU appeared in the study of Mondal [29] and Cay et al. [52]. Mondal used four different solvents (THF, DMF, N,N-dimethylacetamide, and DMSO) and reported that the morphology of the resultant nanofibers changed significantly with the solvent selection. Solvent conductivity and vapor pressure of the solvents were found to be the important factors (**Table 1**). In the study of Cay et al., the effects of the incorporation of ethyl acetate or tetrahydrofuran into TPU/DMF solvent system were investigated. The solutions of TPU in pure DMF and DMF/EA were found to be easily electrospinnable. DMF seemed to be the best solvent to dissolve TPU pellets but diluting TPU solutions with 10 or 20% of EA ensured positive effect on fiber diameter by means of achieving thinner fibers. Incorporation of THF to DMF led to thicker fibers compared to TPU/DMF solutions. With the increasing THF volume fraction, electrospinning is restricted due to high viscosity and low conductivity.

A mixed-solvent system of THF and N,N-dimethylacrylamide (DMAA) in the study of Kidoki et al. to investigate the relationships between the structural features and mechanical properties of electrospun segmented PU (SPU) meshes. They studied the polymer concentration and solvent mixing ratio to achieve different formulations and investigated the operation parameters such as applied voltage, tip to collector distance, and feeding rate. SPU was electrospun from the mixed solvent of THF and DMAA with different mixing ratios [DMF content: 5, 10, and 30% (v/v)]. An increase in DMAA ratio significantly affects the degree of bonding between SPU fibers at contact sites and leads to thinner fibers formation. The porosity of the electrospun SPU meshes decreased with increasing DMF ratio according to the porosimetric characterization. The pore size distribution exhibited three representative peaks of approximately 5, 20, and 70 μm void sizes. Increasing DMAA ratio markedly decreased the proportion of the 20 μm void. In addition to these, an increase in DMAA ratio induced an increase in elasticity of the mesh. The authors pointed that electrospun SPU meshes using a mixed-solvent system with low- and high-boiling point solvents may be useful in the engineering of SPU-fiber-based matrices or scaffolds [57].

4.2. Solution and processing conditions for electrospun PU nanofibers

The diameter of the nanofibers produced by electrospinning is a key parameter for most of the applications. The diameter of the nanofibers defines the structural features such as pore sizes and specific surface areas. These features affect the selectivity of filters, the permeability of filters, catalytic activities in systems using nanofibers to immobilize catalysts, or the cell proliferation in tissue engineering relying on nanofiber-based scaffolds [1]. Solution parameters, especially polymer concentration and the spinning parameters including feeding rate, applied voltage, tip to collector distance, have the strongest impact on the fiber diameter. The properties of PUs can vary in a wide range according to the structure of PU, thus below investigated parameters were particular for the selected PUs.

Solvent	Chemical formula		Molecular weight (g/mol)	Boiling point (°C)	Density (g/mL)	Dipole moment (Debye)	Solubility parameter (δ) $(cal/cm^3)^{1/2}$	Dielectric constant (ε)
N,N-dimethylformamide (DMF)	C_3H_7NO	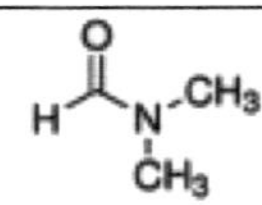	73.09	153	0.944	3.86	12.14	37
Tetrahydrofuran (THF)	C_4H_8O		72.1	65–67	0.889	1.75	9.52	7.5
Ethylacetate (EA)	$CH_3COOCH_2CH_3$	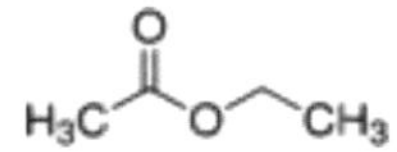	88.11	76.5–77.5	0.902	1.78	9.10	6
N,N-dimethylacetamide	$CH_3CON(CH_3)_2$	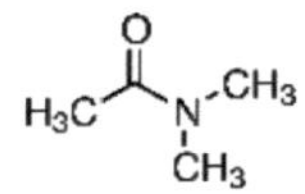	87.12	164.5–166	0.937	3.72	11	37.78
Dimethylsulfoxide (DMSO)	$(CH_3)_2SO$	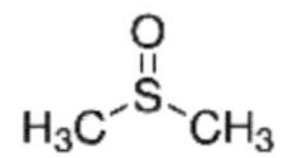	78.13	189	1.10	3.960	12.93	47
Methylpyrrolidone	C_5H_9NO	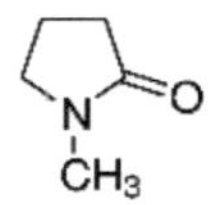	99.13	202	1.028	4.09	10.95	32.2

Table 1. Properties of selected solvents [52–56].

A 3^2 factorial experimental design was used to investigate the tip to collector distance and applied voltage effect for PU nanofibers by Yanilmaz et al. [27]. They used TPU (PU, 270,000 g/mol) supplied from Coim Co. and chose THF for 10% (w/w) TPU solution, 5, 7.5, and 10 cm for tip-to collector distances and 10, 15, and 20 for applied voltages. From their design, it was seen that the distance and applied voltage had significant effects on the fiber diameter and it depends on the applied voltage and distance; furthermore, the interaction of these factors affects fiber diameter significantly.

Akçakoca Kumbasar et al., used DMF for TPU (Pellethane 2103-80AE) solutions and investigated the effect of TPU concentration (6, 8, 10, 12, and 14% (w/w) represented in **Figure 3**), tip to collector distance (8, 10, 12, and 15 represented in **Figure 4**), feeding rate (0.3, 1.5, 2.5, and 3.5 ml/h), and applied voltage (7, 10, 13 kV) on nanofiber diameters. Their results showed that 6% of TPU concentration is too low and 14% is too high for smooth nanofibers and as expected nanofiber diameter increased with increasing TPU concentration. With the increasing tip to collector distance, they were able to achieve better nanofiber morphologies and thinner nanofibers. They showed that decreasing the tip to collector distance caused insufficient solvent evaporation, which makes membrane-like surface instead of nanofibrous surface. Their results also revealed that increase in the feeding rate caused too much polymer deposition on the collector in a short time and that caused conjunction of nanofibers. They concluded that in case of thinner and smoother nanofiber production, 10% of TPU concentration, 20 cm tip to collector distance, 0.3 ml/h feeding rate, and 13 kV applied voltage are suitable [58].

Zhuo et al., also investigated the process parameters, including the applied voltage, feeding rate, and solution concentration. They synthesized PU from PU resin (number-average molecular weight = 180,000 g/mol), based on Poly(e-caprolactone) diol containing a 75% soft-segment content, and 4000 soft-segment length by the bulk polymerization method. They found that 12.0 kV was a critical value for their synthesized PU. For preparing uniform PU nanofibers, diameters ranging from about 700 to 50 nm used 5.0 wt% PU/DMF solutions. However, when the applied voltages were increased to a high value, such as 20.0 or 25.0 kV, the diameters of nanofibers were not uniform and many loops were formed. Higher feeding rate (e.g., 0.1 mm/min) caused larger fibers compared to lower feeding rate (e.g., 0.06 mm/min), and smaller and uniform nanofibers were observed with lower feedings. The fiber diameters increased with the increasing solution concentration. They used five PU/DMF solutions (3.0, 5.0, 7.0, 10.0, and 12.0 wt%), did not observe jet formation in the PU/DMF solutions above 12.0 wt% because of the higher viscosity, whereas in a too diluted solution (e.g., <3.0 wt%), the jet broke into droplets and they observed electrospraying instead of electrospinning. Finally, they concluded that uniform PU nanofibers could be produced by using 5.0–7.0 wt% of PU/DMF solutions, applied voltages of 10–15 kV, and feeding rates of 0.06–0.08 mm/min [38]. They also observed that increasing the applied voltage caused stickier fibers [38].

Andrews et al. investigated some other spinning parameters for Poly(ether urethane), Tecoflex® SG-80A (Thermedics Polymer Products, Wilmington, MA), including flow rate,

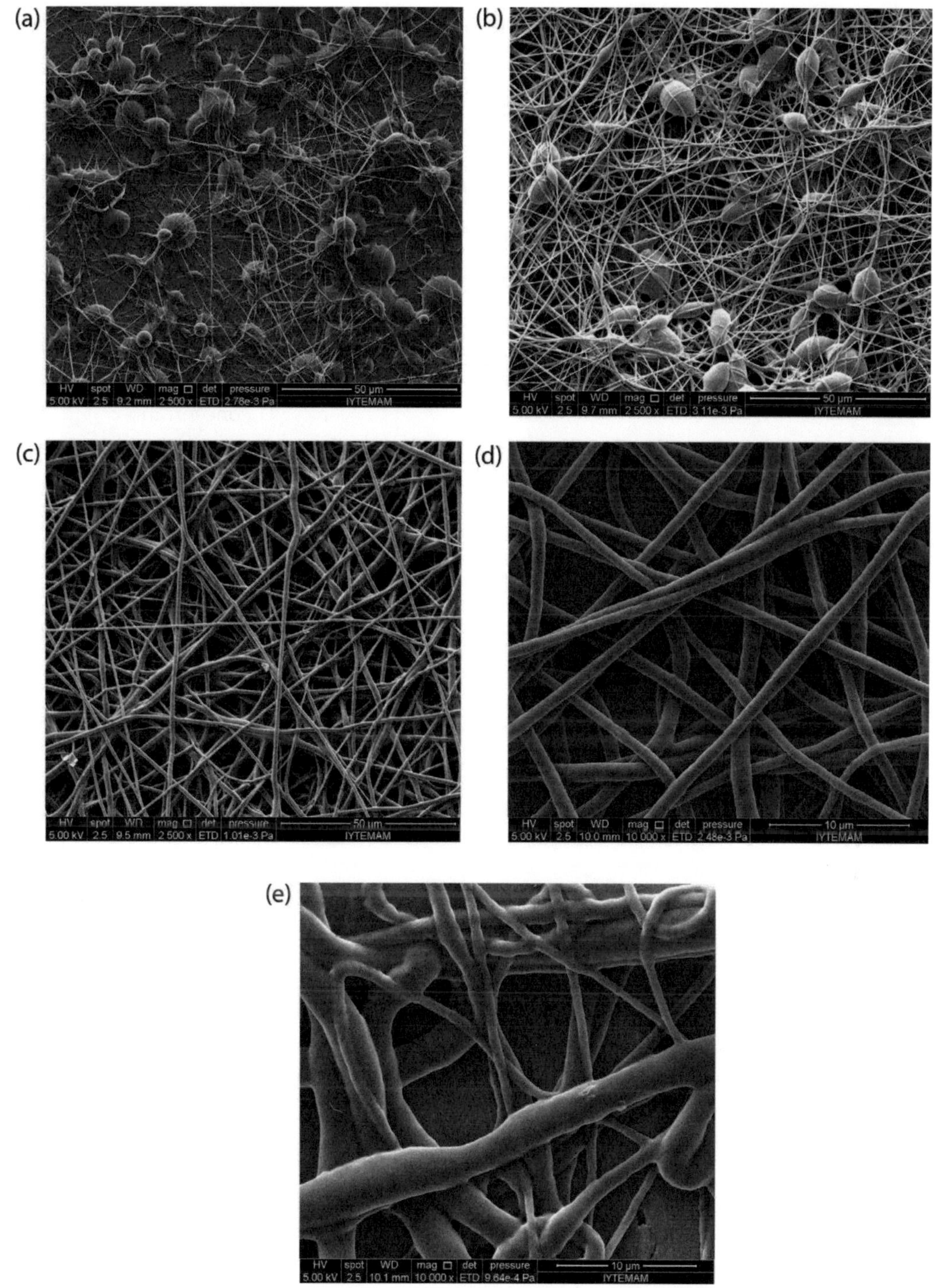

Figure 3. (a) 6%, (b) 8%, (c) 10%, (d) 12%, and (e) 14% w/w electropsun TPU nanofibers [58].

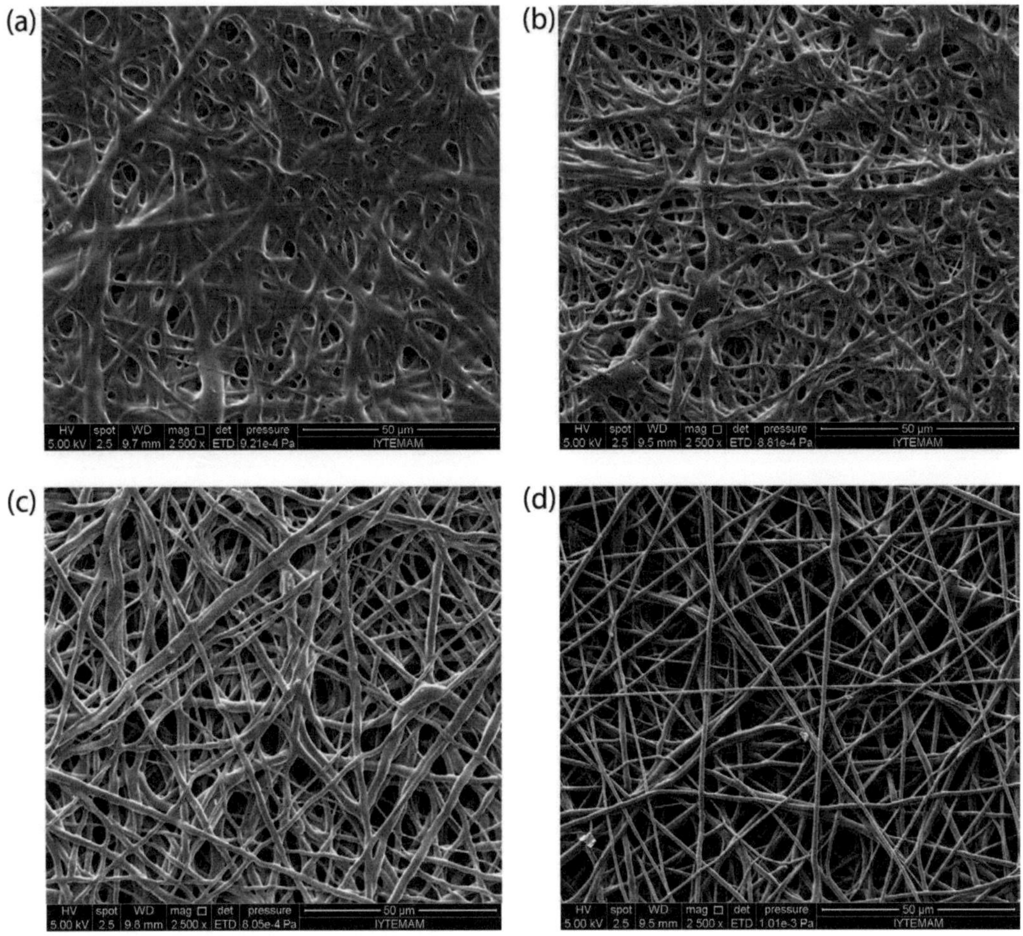

Figure 4. SEM images of the TPU nanofibers at tip-to collector distance: (a) 8, (b) 10, (c) 12, and (d) 15 cm [58].

relative spray height, spray distance, traverse speed, mandrel speed, grid voltage, and mandrel voltage. Dimethylacetamide (DMAc):2-butanone [methyl ethyl ketone (MEK)] were used in the ratio of 1:1.68 to obtain 12.5% w/v PU solution. Their results indicated that inter-fiber separation was significantly affected by flow rate, spray distance, grid voltage, and mandrel voltage, but not by relative spray height, traverse speed, and mandrel speed whereas fiber diameter was significantly affected by flow rate and mandrel voltage, there was no significant difference brought about by changes in relative spray height, spray distance, traverse speed, mandrel speed, and grid voltage. Void fraction was significantly affected by flow rate but not by relative spray height, spray distance, traverse speed, mandrel speed, grid voltage, and mandrel voltage. Fiber orientation on the external surfaces was significantly affected by traverse speed and mandrel speed, but not by flow rate, relative spray height, spray distance, grid voltage, and mandrel voltage. They indicated that the volumetric flow rate was the sole spinning parameter that affects the scaffold thickness [59].

Demir et al. prepared a segmented polyurethaneurea based on poly(tetramethylene oxide)glycol cylcoaliphatic diisocyanate, and unsymmetrical diamine. They used 2-methyl-1,5 diaminopentane (DAP), dibutylamine (DBA), DMF as solvents and studied electrospinning behavior of produced elastomeric polyurethaneurea copolymer in solution. They observed fiber diameters increased with the increasing solution concentration and lower concentration favored beads, and increased concentrations favored curly fibers. Salt addition increased the solution conductivity, which led to increase in mass flow. They found viscosity and solution temperatures were dominant factors and improving the fiber morphology was possible with increasing solution temperature, and it was quicker to electrospun these solutions compared to the solutions that were at room temperature [20].

The effect of tetraethylammonium bromide (TEAB) salt on the spinnability of polyurethane (PUR, Larithane LS 1086, aliphatic elastomer based on 2000 g/mol, linear polycarbonated diol, isophorone diisocyanate, and extended isophorone diamine) was investigated in the study of Cengiz et al. They used a roller electrospinning method. They found that the conductivity, viscosity, spinning performance increased with salt concentration. Also, solution viscosity decreased with shear rate. PU including 1.82 wt% TEAB gives the best spinning performance although 0.87 wt% TEAB is the optimum value related to fiber properties such as diameter, uniformity, and morphology given the ideal PU nanoweb structure [60]. In another study of Cengiz et al., they discussed the effects of 1,1,2,2 tetrachlorethylen (TCE), a non-solvent addition on the independent (electrical conductivity, dielectric constant, surface tension, and the rheological properties of the solution etc.) and dependent parameters (number of Taylor cones per square meter (NTC/m^2), spinning performance for one Taylor cone (SP/TC), total spinning performance (SP), fiber properties such as diameter, diameter uniformity, non-fibrous area). The effect of non-solvent concentration on the dielectric constant, surface tension, rheological properties of the solution, and also spinning performance were statistically important. Beside, non solvent concentration affects the quality of fiber and nanoweb structure [61].

Yalcinkaya et al., measured the jet current and jet life in roller electrospinning of PU (molecular weight 2000 g/mol, Larithane LS 1086; Novotex, Italy) in their study. They analyzed the relationships between jet current and jet life and number of Taylor cones/m^2 (NTC/m^2), spinning performance (SP), and fiber properties (diameter, non-fibrous area) and determined the effects of PU and TEAB concentrations on jet current and jet life. They observed that jet current increases with PU and TEAB concentration, while jet life decreases. NTC/m^2 and spinning performance increased with jet current and decreased with jet life, and they observed that jet current movement gives an idea about jet life [62].

Among above the traditional electrospinning method, an uncommon laser-heated electrospinning which is represented in (**Figure 5**), was used by Takasaki et al. and they investigated the effect of the spinning conditions including the applied voltage, the laser power, the laser irradiation point, and the laser beam width on the diameter of TPU microfibers. The average diameter of electrospun TPU fibers decreased with decreasing applied voltage and increasing laser power. A narrower laser beam reduced the variation in the fiber diameters. A PU microfiber with an average diameter of 2.4 μm and a coefficient of variation of 8% was obtained using a 0.9 mm wide laser beam [63].

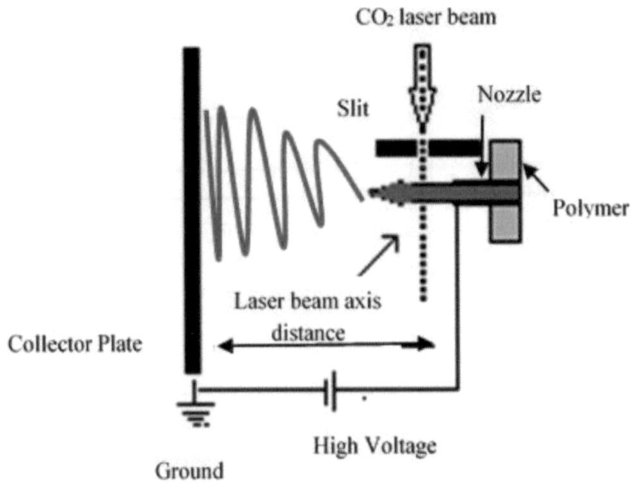

Figure 5. Schematic diagram of laser-electrospinning system. Modified from Takasaki et al. [63].

4.2.1. Melt electrospinning

Although most electrospinning researches are based on polymer solutions, there are also several researchers that use polymer melts. Generally, most electrospinning conditions also affect the molten polymer in electrospinning [9]. Melt electrospinning is especially attractive for tissue engineering scaffold manufacturing process, because it is possible to eliminate potentially cytotoxic solvents in contrast to electrospinning from a polymer solution. However, there are some technical challenges related to the need for a well-controlled high-temperature setup and there is a difficulty in developing an appropriate polymer. In the study of Karchin et al., a biodegradable and thermally stable PU, produced from 1,4-butanediamine and 1,4-butanediol, $(CH_2)_4$-content diisocyanates and polycaprolactone. These aliphatic PU formulations were used in the melt electrospinning [42]. A catalyst-purified PU based on 1,4-butane diisocyanate, polycaprolactone, and 1,4-butanediol in a 4/1/3 M ratio respectively, yielded a non-toxic polymer that could be electrospun from the melt. The authors concluded that this electrospun polymer contained point bonds between fibers and its mechanical properties were analogous in vivo soft tissues.

4.3. Collectors for electropsun PU nanofibers

In most electrospinning setup, the collector plate is made out of a conductive material and covered by a piece of aluminum foil, which is electrically grounded so that there is a stable potential difference between the source and the collector [9]. Rotating drum with different diameters [47, 59, 64–69], parallel electrodes [70], rotating wire drum collector [71], rotating tube collector with knife-edge electrodes below [66], disc collector [72–74] array of counter-electrodes [75], rotating drum with sharp pin inside [76], ring collector placed in parallel [77] could be also used for different purposes [13]. Despite the random and centered collection of nanofibers on the collector plate, the most basic form of getting aligned nanofiber deposition is through the use of a rotating mandrel. Schematic representation is given in **Figure 6**. Beside aligned nanofibers, the texture of the fiber mesh may also be varied by using patterned collectors.

Andrews et al. deposited the PU nanofibers onto 3 mm diameter stainless steel mandrels that were pre-coated with a saturated NaCl solution to facilitate fiber formation and subsequent removal of the scaffold [59]. Pedicini and Farris used a grounded flat aluminum foil target for isotropic TPU fiber mats for tensile tests and infrared spectroscopy experiments, oriented electrospun TPU nanofiber samples for IR dichroism studies were collected onto a rotating stainless steel drum [64]. Two types of electrodes with tines were used in the study of Banuskeviciute et al. [79]. The electrode consists of eight separate plates. In every plate, tines were set at equal distances. Tines were different by the width and shape of every electrode. They showed that the type of electrode had an influence on the structure of the electrospun TPU mats, but not on the diameter of formed fibers. Rotating mandrels could be used for electrospun vascular graft for the regeneration of blood vessel (**Figure 6b**). Theron et al. [65] spun small diameter vascular graft prototypes (1.6 mm nominal ID) using the apparatus including rotating/translating mandrel. Tubes were removed from the mandrels by swelling in EtOH and dried. Thandavamoorthy et al. [80] self-assemble electrospun PU nanofibers into honeycomb patterns on the collector surface. Residual charges on the collected fibers and the electrical property of the collector screen influenced the self-alignment of fibers. They electrospun PU nanofibers over these substrates while keeping all other process parameters constant. When cotton, a natural fiber with very poor electrical conducting properties, was used as the collecting surface, a 3-D honeycomb pattern deposition was produced (**Figure 7**). Similar results were observed with the glass substrate.

4.4. Electrospinning of PU/blends and PU nanocomposite nanofibers

In electrospinning, it is sometimes useful to combine the properties of two or more polymers to achieve a new structure. This can be obtained either by physical mixing to form a blend or through polymerization to form a copolymer [9]. Addition of a second component could also facilitate the electrospinning process. For example, it was difficult for most of the natural polymers to be electrospun; however, addition of a synthetic polymer could improve the

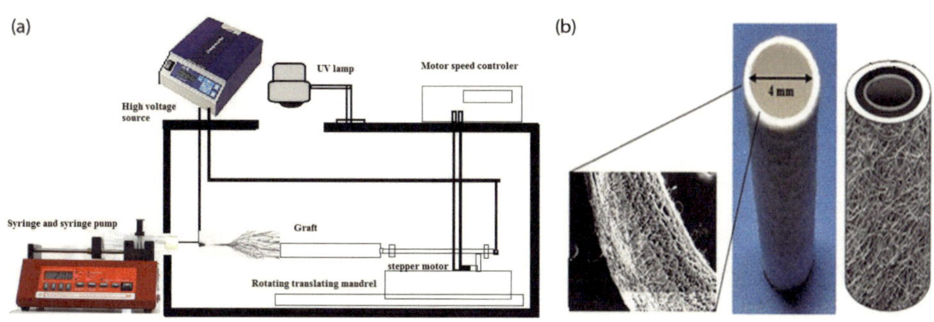

Figure 6. (a) Schematic representation of the reactive electrospinning apparatus with UV light using a rotating mandrel. Modified from Theron et al. [65]. (b) Electrospun vascular graft for regeneration of blood vessel [78].

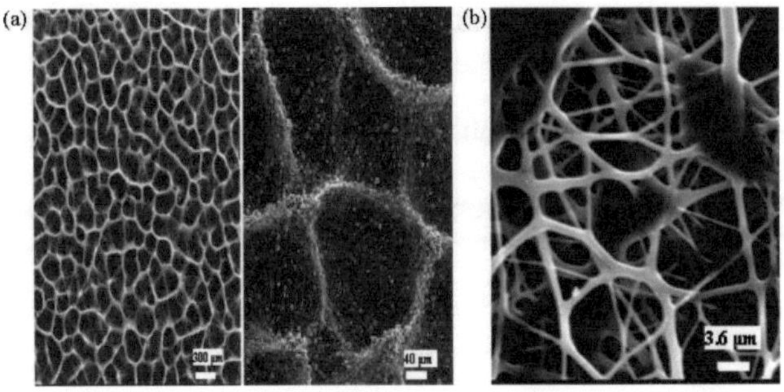

Figure 7. SEM images of self-assembled polyurethane electrospun nanofibers [81] .

processability of these polymers [82]. PUs are easy to electrospin and they can be mixed either with a natural polymer or with a synthetic polymer [83, 84, 87] for special applications such as collagen [88], dextran [43], and hydroxypropyl cellulose (HPC) [41].

Buruaga et al. [83] dissolved a water-soluble polymer [poly(ethylene oxide) (PEO)] in the PU dispersion and fibers were obtained from electrospinning of the resulting mixture. The template polymer (PEO) was removed from electrospun fibers by water extraction, then they obtained pure PU fibers. The authors offered a new perspective for the preparation of micro and nanofibers by using aqueous dispersions for the preparation of water-insoluble PU fibers by electrospinning. Lee et al. [84] dissolved various polyblends of poly(vinyl chloride) and PU (Pellethane 2363-80AE) in a mixture of THF and DMF. They produced nanofibers in different ratios, with several electrospinning conditions and investigated the relationship between morphology and mechanical behavior of the resulting fiber mats. Point-bonded structures in the PU fiber mats increased with increasing PU composition and the mechanical properties of the fiber mats. In the study of Hong et al. [44], PU/organically modified MMT (O-MMT) nanocomposites were prepared via a solution intercalation method and electrospun. The authors investigated the effect of O-MMT on the morphology and physical properties of the PU/O-MMT nanofiber mats. To prepare the PU and PU/O-MMT nanofibers, 11 wt% of PU and PU/O-MMT solutions in a mixed solvent of DMAc/THF (7/3 w/w) were electrospun. Increasing the content of O-MMT resulted in the linearly increase of conductivities of the PU/O-MMT solutions and this decreased the average diameters of the PU/O-MMT nanofibers. Produced PU and PU/O-MMT nanofibers were uniform and not microphase separated. They achieved a well distributed and oriented MMT layers within the PU/O-MMT nanofibers. When the PU/O-MMT nanofibers were annealed, the exfoliated MMT layers hindered the microphase separation of the PU. Incorporation of MMT layers into PU nanofibers improved Young's modulus and tensile strength of PU/O-MMT nanocomposites. Akçakoca Kumbasar et al. [85] loaded β-CD into TPU nanofibers (**Figure 8**). They observed that TPU/CD nanofibers had higher fiber diameters compared to pure TPU nanofibers and fiber diameters increased with the increase in β-CD concentration. The authors also proved the inclusion complex formation capability of TPU/CD nanofibers by the phenolphthalein test method.

Hu and Yu [86] prepared shape stabilized bio-phase change material (PCM) by encapsulating the wax inside the PU nanofibers using coaxial electrospinning. The encapsulated bio-PCMs

Figure 8. 30% β-CD loaded (a) 8% TPU, (b) 10% TPU nanofibers.

can potentially be used for thermal storage and thermal protection areas, which has appealing environmental advantages. The authors encapsulated the soy wax into PU fibers without being miscible with PU fibers and adjusted the wax content either by the concentration of wax/chloroform solution or flowing rate. Results of thermal analysis showed that the enthalpy increased as the wax content increase. The authors conducted 100 heating-cooling cycles and the thermal properties of the fibers were unaltered. The outer PU layer prevented the leakage of the bio-wax like a reservoir and it also enhanced the modulus and lowered the tensile strain. They produced uniform fiber morphology with a core-shell structure and a homogeneous wax distribution throughout the core of the fibers.

Vlad et al. [41] synthesized PU from hexamethylene diisocyanate (HDI), polytetramethylene ether glycol (PTMEG), and butanediol (BD). They mixed PU with different proportion of hydroxypropyl cellulose (HPC) for biomedical applications. Increase of HPC amount of sample provoked a decrease of contact angles. Similar behavior was observed for fibrinogen adsorption, which confirms that PU/HPC nanofibers are suitable in biomedical applications. Unnithan et al. [87] combined PU with two biopolymers, cellulose acetate (CA) and zein to produce an antibacterial electrospun nanofibrous scaffolds. In another study, they used a solution composed of dextran, PU, and ciprofloxacin HCl (CipHCl) drug for wound dressing applications [43]. They investigated the viability, proliferation, and attachment of fibroblasts to the PU-dextran and PU-dextran-drug scaffolds. Their results indicated that the composite mat has a good bactericidal activity and the cells especially interacted with the drug containing scaffolds.

Chen et al. [88] produced collagen functionalized-TPU nanofibers (TPU/collagen) by coaxial electrospinning technique (**Figure 9**) with a goal to develop biomedical scaffold. 1,1,1,3,3,3-hexafluoro-2-propanol (HFP) was used as solvent for collagen and TPU. The authors carried out the cross-linking process by placing electrospun membrane (collagen (shell)/TPU (core) (8 wt%/3 wt%) together with a supporting aluminum foil in a desiccator using glutaraldehyde (GTA) (25% water solution) with different process time. Feasibility of PU/collagen core-shell construct as an optimal tissue engineering scaffold materials was supported by its high porosity and adequate pore size. Pig illiac endothelial cells proliferation in vitro demonstrated the feasibility and efficacy of using TPU/collagen composite nanofibers for improving cell-scaffold

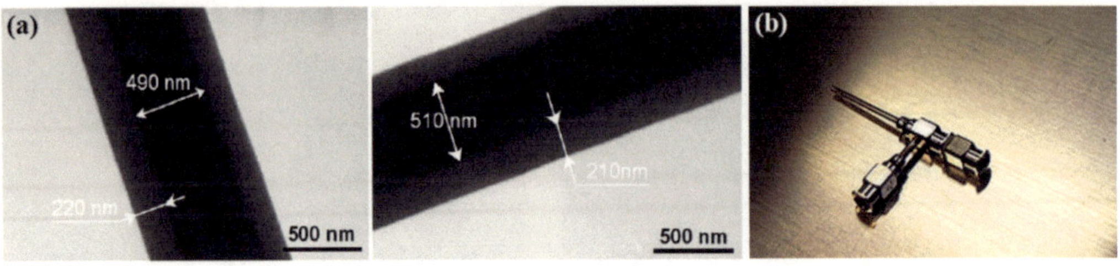

Figure 9. (a) Transmission electron microscopy image of the core shell nanofibers, (b) coaxial nozzle for electrospinning [89].

interactions pore size and these composite nanofibers had the characters of native extracellular matrix and may be used effectively as an alternative material for tissue engineering and functional biomaterials.

Huang et al. [90] electrospun random and aligned nanofibrous scaffolds based on collagen-chitosan-TPU blend to mimic the componential and structural aspects of the native extracellular matrix. They also investigated the optimal proportion to keep the balance between biocompatibility and mechanical strength. The scaffolds were cross-linked by GTA vapor to prevent them from being dissolved in the culture medium. The mechanical properties of the scaffolds were found to be flexible with a high tensile strength. Cell viability studies with endothelial cells and Schwann cells demonstrated that the blended nanofibrous scaffolds had good biocompatibility and aligned fibers could regulate cell morphology by inducing cell orientation. Chen and Chiang grafted collagen to PU fiber surface by low temperature oxygen plasma treatment, which could improve surface hydrophilicity to promote wound healing and facilitate covalent binding of collagen molecules to the plasma-treated PU surface [91]. After modification, the nanofibrous membrane's antimicrobial activity improved to ~100% inhibition of bacterial growth. Water absorption ability of membrane was increased, which facilitates its use as a functional wound dressing. Also, their results demonstrated that the nanofibrous membrane was better than gauze and commercial collagen sponge wound dressing in wound healing rate.

4.5. Mechanical properties of electrospun PU nanofibers

The electrospinning technique have the advantages of being simple, convenient, and inexpensive in comparison with conventional methods such as wet, dry, and melt spinning. Unfortunately, the practical applications have been limited because produced electrospun fiber mats have poor mechanical properties, low molecular orientation, and broad distribution of fiber diameter. Hence, an enhancement in both mechanical and physical properties of the electrospun fiber mats is very important from an industrial point of view [92]. TPUs present a class of polymers that possess a range of very desirable properties: they are elastomeric, resistant to microorganisms and abrasion, and have excellent hydrolytic stability and many commercially available TPUs can be used to make good electrospinning solutions [64]. Earlier studies on the electrospinning process have been more focused on the basic principles [2–4] and processing parameters like the voltage applied, tip-to collector distance, and viscosity of solution [20, 26–28, 48]. The future use of electrospun PU materials in practical applications will require good mechanical properties. Thus, several authors recently investigated the mechanical properties of electrospun PU fiber mats [19, 40, 64, 65, 92–94].

Lee et al. [92] used Pellethane 2363-80AE and solved it in a mixture of THF and DMF (60/40, v/v) at room temperature at a concentration of 8 wt%. The authors investigated the mechanical behaviors of TPU by cyclic tensile tests. Produced electrospun TPU mats were composed of randomly oriented sub-micron fibers, where each fiber was restricted by physical netting and entanglements. They have seen almost linear elastic behavior until the fiber mats undergo breaking. Major cause of energy loss and stress softening at relatively low strains was the slippage of the electrospun fibers. At higher strains, the breaking of electropsun fiber at point-bonding junctions, as well as the slippage crossed fibers occurred as a further source of the dissipation energy. Pedicini and Farris [64] prepared the Pellethane 2103-80AE in DMF at room temperature at a concentration of 7% by weight and bulk samples by thermally processing plaques from polymer pellets. The uniaxial tensile tests results indicated that the behavior of the electrospun Pellethane 2103-80AE to be distinctly different from the bulk. Qualitatively, the electrospun mat is also elastomeric in nature, but the shape of the stress-strain curve for the electrospun material is not sigmoidal in contrast to bulk material. The curve is monotonic and its slope has not got an inflection. When they applied a strain to the electrospun mat, they got oriented fibers. They also concluded that the apparent molecular orientation in the electrospun fibers leads to the pronounced reduction in elongation to failure of the electrospun mat, relative to the bulk.

Cha et al. [93] synthesized shape-memory PU block copolymers to prepare electrospun non-wovens. The authors prepared PU solutions in a mixture of DMF and THF, and electrospun PU non-wovens with hard-segment concentrations of 40 and 50 wt%. The average diameter of low viscosity (ca. 130–180 cPs) beaded electrospun fibers was about 800 nm. In contrast, the average diameter of high viscosity (ca. 530–570 cPs) electrospun fibers was about 1300 nm. The mechanical properties of the electrospun PU non-wovens were investigated and found the increase in the hard-segment concentration increased the tensile strengths as well as the viscosities. Also, because of a difference in the velocities of the drum collectors, the tensile strength in the machine direction was higher than that in the transverse direction. Prepared PU non-wovens have a shape recovery of more than 80% that included hard-segment concentrations of 40 and 50 wt%.

The main weakness of electrospun nanofibrous membrane structures seems to be their poor mechanical properties caused by relaxation processes occurring immediately after fiber formation, at which a certain degree of molecular orientation is lost. With the aim to overcome this problem, some researchers combined single-walled carbon nanotubes (SWCNTs) or multi-walled carbon nanotubes (MWCNTs) (**Figure 10a**) [94] to prepare PU nanocomposite fibers and achieve a significantly enhanced Young's modulus [19, 40, 95].

Sen et al. [19] demonstrated the effect of the chemical functionalization of SWNTs on the mechanical properties of SWNT-reinforced composites of electrospun PU nanofibers. The tensile strength of ester functionalized and as-prepared SWNT-PU membranes is enhanced

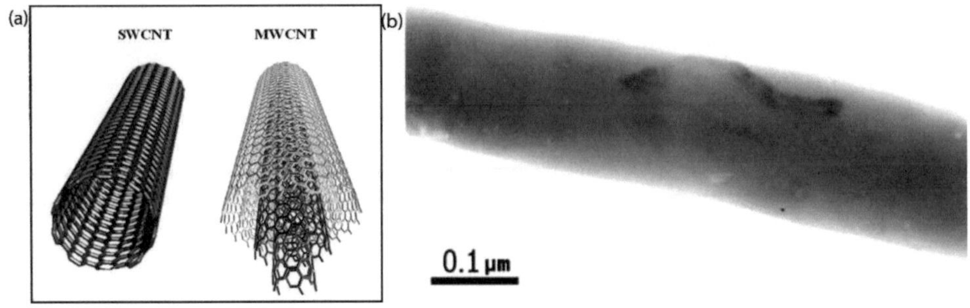

Figure 10. (a) TEM images of MWNTs/PU nanofibers and multi-wall carbon nanotube (MWCNT) [94]. (b) Schematic diagrams of single-wall carbon nanotube (SWCNT) [95].

by 104 and 46%, respectively as compared to electrospun pure polyurethane membranes. The tangent moduli of as-prepared and ester functionalized SWNT-PU membranes were also higher than the control PU membranes and were 215 and 250%, respectively.

Liu and Pan [95] produced MWNTs/PU composite nanofibrous membranes (**Figure 10b**). They used a thermal treatment method to improve the comprehensive properties of the membranes. They reported that the proposed thermal treatment method effectively improved the conductivity and mechanical properties of the nanomaterials, with 3 wt% of MWNTs, the conductivity of the membranes reached 8.29×10^{-5} Scm^{-1}, which was nearly three orders higher than that of the untreated sample and the tensile strength and the modulus increased more than 50% after thermal treatment. PU and PU/multi-walled carbon nanotube (MWCNT) nanocomposite nanofibers, both with diameters of 350 nm, were prepared by Kimmer et al. as well [40]. The appearance of smaller nanowebs in PU/MWCNT nanofiber structures with diameters of 20–40 nm was observed. They attributed the existence of these structures to the occurrence of strong secondary electric fields, which were created between individual conducting MWCNTs (distributed in the PU/MWCNT nanocomposites). From the TEM images individual and very well aligned MWCNTs can be seen in the outer surface of the main nanofiber from which nanoweb fiber was created. However, the authors did not investigate the mechanical properties of the prepared electrospun fiber mats.

Cross-linking the nanofibrous mats is another way to improve the mechanical properties. Theron et al. [65] modified a medical standard TPU (Pellethane 80A) with latent cross-linkable groups; and they determined the effect of subsequent cross-linking on viscoelastic properties and degradation resistance. They confirmed the successful cross-linking by insolubility of the materials. Pellethane was cross-linked with containing an increasing number of pentenoyl groups. Cross-linking decreased (up to 42%, $p < 0.01$) the hysteresis and creep (44%, $p < 0.05$), and significantly improved the degradation resistance in vitro. Modified Pellethane was also electrospun into tubular grafts and with UV irradiation they were cross-linked to make them insoluble. Prototype grafts had a burst pressure of >550 mmHg, and for uncross-linked and cross-linked samples, it was 12.1 ± 0.8 and 6.2 ± 0.3%/100 mmHg, respectively. They concluded that the modification of Pellethane with PCL occurred on the carbamate nitrogen and cross-linking resulted in improved viscoelastic properties. Although there were some decrease in tensile strength and strain, the polymers had sufficient strength and extensibility. They suggested to use these cross-linked materials such as vascular grafts where repetitive and relatively low stresses is encountered.

5. Potential applications of electrospun PU nanofibers

PU nanofibers prepared by electrospinning technology may have almost circular cross-sections, smooth surfaces, and diameters ranging from a few nanometers to several micrometers. Electrospun nanofibers have great applicability for drug delivery applications [16, 96–101], medical implants [102], nanocomposites for dental restoration [103], preservation of bioactive agents [104, 105], tissue engineering [12, 106], wound dressing [21, 43, 82], biosensors [107, 108], molecular separation [109], filters [45, 110, 111], and protective clothing [112–114]. Potential applications of electrospun PU nanofibers are discussed in depth below.

5.1. Biomedical applications

The importance of electrospinning, in general, for biomedical applications like wound dressing, drug release, tissue engineering, medical implants etc. is emphasized in this part of the chapter. The focus is on the active substances combined with PU and blends of PU that have been electrospun and also on the modifications that have been carried out in conventional electrospinning apparatus.

PUs, due to their structure/property diversity, are considered one of the most bio- and blood-compatible materials known today. Properties like durability, fatigue resistance in tensile, compression or shear, elasticity, compliance, "elastomer character," and propensity for healing became attainable via PU. Furthermore, modification via hydrophilic/hydrophobic balance or by attachments of biologically active species to the PUs is possible [37].

Due to their higher surface area to volume ratio, nanofiber mats have been studied for applications as drug delivery carriers. The blending (or mixing) technique is a common choice for the nanofiber functionalization [9]. Verreck et al. [96] prepared PU nanofibers containing model drug itraconazole and ketanserin, using DMF and dimethylacetamide (DMAc) as solvent, respectively. The collected non-woven fabrics released the drugs at various rates and profiles based on the nanofiber morphology and drug content. They used a specially designed release apparatus based around a rotating cylinder for release studies. They loaded 10% ketanserin and 10 and 40% itraconazole to the PU nanofibers. At low drug loading, itraconazole was released from the nanofibers as a linear function of the square root of time. They did not observe initial burst release. They explained the faster initial release of ketanserin versus itraconazole with (1) higher drug solubility of the ketanserin in the polymer and (2) increased drug diffusivity in the polymer. They also observed that the obtained fiber diameters did not significantly influence the initial release rate. Fiber diameters for the ketanserin loaded samples were between 0.5 and 2 μm, 10% itraconazole loaded were 2 μm and 40% loaded samples were 0.3–0.7 μm. They observed a biphasic release pattern for ketanserin in which two sequential linear components. The authors correlated these release phases temporally with (1) drug diffusion through the polymer and (2) drug diffusion through formed aqueous pores.

Akduman et al. [99] prepared nanofiber mats of TPU containing naproxen (NAP) from 8 to 10% (w/w) TPU/DMF solutions. The amount of NAP in the solutions was 10 and 20% based on the weight of TPU and the collection periods were changed to 5, 10, and 20 h. The diameters of the nanofibers were significantly affected by the TPU concentration; however, the NAP loading and percentage of NAP did not have a significant effect on the fiber diameters. They investigated the release characteristics of fiber mats by the total immersion method in the phosphate buffer solution at 37°C and performed the characterization of the produced NAP-loaded TPU mats. They observed that the diffusion paths correlated to the nanofiber collection period has a significant effect on the release characteristics of the drug. Short collection periods resulted burst release of the drug. Beside, higher drug loading (20% $w/w_{polymer}$) caused higher drug release rate. The drug that had previously leached from the nanofiber mat formed channels through the matrix and that these channels lead to higher release rates of the drug. As a result, the authors suggested that in drug-loaded electrospinning studies, the produced mats should be collected for at least 20 h for a one needle electrospinning system, and drug loading should not exceed 10% for better controlled release rates (**Figure 11**).

Polymeric nanofiber matrix has similar structure with the nano-scaled non-woven fibrous extra cellular matrix (ECM) proteins, thus it can be considered as potential candidate for ECM-mimetic material [115]. A successful tissue engineering nanofibrous material should allow cell attachment and proliferation. Most of the surface modifications are related to the biocompatibilities of polymeric tissue engineering scaffold and to the immobilizations of biomolecules that can be specifically recognized by cells on the biomaterials [9]. In the study of Unnithan et al. [43] prepared an antibacterial electrospun scaffold by electrospinning of a solution composed of dextran, polyurethane (PU), and ciprofloxacin HCl (CipHCl) drug. They used dextran, which is a versatile biomacromolecule. Dextran can be used by blending with either water-soluble bioactive agents or hydrophobic biodegradable polymers for biomedical applications for preparing electrospun nanofibrous membranes. They investigated the interaction

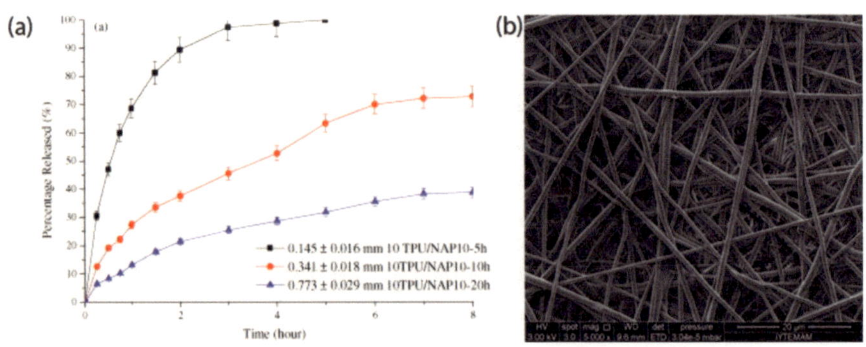

Figure 11. (a) Percentage release of NAP from (■) 5 h collected-, (•) 10 h collected , and (▲) 20 h collected 10TPU/NAP10 mats by the total immersion technique during (b) SEM image of a NAP-loaded TPU nanofibers [99].

parameters between fibroblasts and the PU-dextran and PU-dextran-drug scaffolds such as viability, proliferation, and attachment. Their results indicated that the cells interacted favorably with the scaffolds especially the drug containing one. The authors concluded that the introduced scaffold might be an ideal biomaterial for wound dressing applications.

Amoroso et al. [116] utilized two fabrication modalities to induce controlled alterations in fiber network topology. They investigated the variation of collecting mandrel translation velocity, and concurrent electrospraying of cell culture medium with or without cells or rigid particulates to emulate the maximum possible micro-inclusion stiffness. Once, they electrospun poly(ester urethane) urea (PEUU) without modifications to the process, then electrospun it "wet" by concurrently electrospraying cell culture medium onto the target. The authors studied the effect of cell and particulate inclusion into the fiber scaffold matrix. Authors used either vascular smooth muscle cells or polystyrene microspheres. They electrosprayed the cells at concentrations of 2 and 6 million/mL and electrosprayed the microspheres at 7 million/mL into the cell culture medium. The inclusion of cell culture medium into the construct resulted in a dramatic change in scaffold microarchitecture. Particulates and cells would act as additional fiber bonding sites, increasing the effective fiber intersection density and consequently raising the level of mechanical anisotropy. Practically, wet processing and mandrel rastering can be successfully implemented as tools to reliably modify scaffold microarchitecture without altering fiber alignment.

In the study of Carlberg et al. [117], electrospun fibrous PU scaffolds have been studied as substrate for embryonic stem cells cultivation and neuronal differentiation. The authors prepared electrospun scaffolds composed of biocompatible PU resin (Desmopan 9370A) with a vertical electrospinning setup. They showed that the embryonic stem cells displayed favorable interaction with the substrate, spreading outgrowths, establishing connections to adjacent cells and attaching to individual fibers. Immunocytochemistry results showed that fibers can support neuronal differentiation in embryonic stem cell cultures. Their results indicated that physical cues induced by the fibrous scaffolds affect stem cells toward a neuronal cell fate. Hence, they claimed that these scaffolds are potential cell carriers in neural tissue engineering repair and rehabilitation of the adult human nervous system.

Grasl et al. [118] in vitro studied the mechanical homogeneity of electrospun small diameter polyurethane grafts as well as spontaneous attachment, proliferation, and adhesion molecule expression of endothelial cells. They prepared the prostheses from 5% (w/w) PU (Pellethane 2363-80A; DOW Plastics) in 1,1,1,3,3,3-hexafluoro-2-propanol and used a mandrel with the diameter of 2.1 mm and the length of 170 mm. It was rotated at 200 rpm and oscillated 150 mm in the transverse direction at a speed of 8 mm/s. The authors measured the axial and circumferential tensile strengths and they found that they were two-fold higher in the circumferential direction. They fabricated highly uniform small diameter polyurethane grafts and easily achieved the endothelial cells attachment without precoating the fiber matrix. They also observed that the synthetic graft surface neither impaired the endothelial response toward IL-1b stimulation nor did it adversely affect the regulation of expression of endothelial adhesion molecules.

To assess mesh architecture sensitivity to manufacturing parameters, Mitchell and Sanders [119] developed a system for controlled electrospinning of fibro-porous scaffolds for tissue engineering applications. Their intent was to achieve scaffolds with well-controlled fiber diameters and inter-fiber spacing. They used a custom, closed-loop controlled, electrospinning system. With their system, they were capable of producing TPU meshes with fiber diameters ranging from 5 to 18 m with variability less than 1.8%; inter-fiber spacing ranged from 4 to 90 m with variability less than 20.2%. They concluded that their system has potential use in biomedical applications where meshes with controlled fiber diameter and inter-fiber spacing are needed.

Antibacterial and antimicrobial agents such as silver (Ag) nanoparticles [120], 4-vinylpyridine (4VP) [36], or streptomycin sulfate [87] loaded electrospun PU nanofibers were also developed for biomedical applications. Some of these studies combined the antibacterial properties of PU nanofibrous membranes with wound dressing applications.

Yao et al. [36] developed a novel antibacterial material by surface modification of electrospun PU fibrous membranes, using a plasma pretreatment, UV-induced graft copolymerization of 4-vinylpyridine (4VP), and quaternization of the grafted pyridine groups with hexylbromide. Poly(4-vinyl-*N*-hexyl pyridinium bromide) was grafted to the surfaces to achieve antibacterial activities. They showed that the morphologies of PU fibrous membranes changed slightly during the modification process. The tensile strength of PU fibrous membranes decreased after surface modification. After the modification, the tensile strength of PU fibrous membranes from 7% (w/v) decreased from 3.27 to 1.99 MPa, losing almost 40% of tensile strength. They observed smaller decreases in the tensile strength with the increasing solution concentration. The largest diameter of fibers belongs to a concentration of 11% (w/v) and the loss of tensile strength was approximately 16%. They carried out the antibacterial assays with surface modified PU fibrous membranes electrospun from 10% (w/v). Their modified PU fibrous membranes possessed highly effective antibacterial activities against Gram-positive *Staphylococcus aureus* (*S. aureus*) and Gram-negative *Escherichia coli* (*E. coli*) and the authors claimed that these fibrous membranes may have a wide variety of potential applications in high-performance filters, protective textiles, and biomedical devices.

Sheikh et al. [120] synthesized PU nanofibers containing silver (Ag) nanoparticles. They carried out the synthesis of silver nanoparticles by exploiting the reduction ability of DMF, which is used mainly to decompose silver nitrate to silver nanoparticles. Typically, a sol-gel consisting of $AgNO_3$/PU was electrospun and aged for 1 week. Ag nanoparticles were created in/on PU nanofibers. They examined the durability of the silver NPs on the PU nanofibers by harsh successive washing. Their results confirmed the good stability of the nanofiber mats. The authors used *E. coli* and *Salmonella typhimurium* to check the antimicrobial influence. Consequently, antimicrobial tests indicated that the prepared nanofibers have a high bactericidal effect and they have potential for using as antimicrobial agents.

In the study of Unnithan et al. [87], an antibacterial electrospun nanofibrous scaffolds with diameters around 400–700 nm were prepared by physically blending PU with two biopolymers, cellulose acetate (CA) and zein. They used PU as the foundation polymer, blended it with CA and zein to achieve better hydrophilicity, cell attachment, proliferation and blood clotting ability. They incorporated an antimicrobial agent, streptomycin sulfate into the electrospun fibers and characterized the interaction between fibroblasts and the PU-CA and

PU-CA-zein-drug scaffolds. They investigated the viability, proliferation, and attachment of the fibroblasts on the nanofiber scaffolds and observed that the produced composite nanoscaffold has better blood clotting ability than pristine PU nanofibers. They found that the incorporation of CA and zein to the nanofiber membrane enhanced the bioactivity of nanofiber mats, as weel as the hydrophilicity. CA and zein also provided a moist environment for the wound.

Nanofibrous materials provide a realistic representation of the native tissue than any other substrate with respect to general cell culture. An advantage of using nanofibrous material is they can also be produced in a highly aligned orientation. The orientation of the nanofibers plays an important role in the study of cell behavior whose native environments consist of highly aligned ECM [121]. Most researchers carried out cell culture tests to characterize the developed nanofibers for wound dressing and scaffold purposes. In **Figure 12**, representative cell-cultured images of nanofibers were given.

Khil et al. [21] prepared PU nanofibrous membrane and evaluated its performance as a wound dressing. They saw that the produced nanofibrous wound dressing showed controlled evaporative water loss, good oxygen permeability, and promoted fluid drainage ability due to porosity of nanofibrous membrane. Neither toxicity nor permeability to exogenous microorganism was observed with the nanofibrous membrane. They also observed that the epithelialization rate was increased confirmed by histological examination, and they were able to control the exudate in the dermis by covering the wound with the electrospun membrane. Thus, they claimed that nanofibrous PU membrane prepared by electrospinning could be properly employed as wound dressings.

Kim et al. [39] prepared a blended nanofiber scaffold using synthetic and natural polymers, PU and gelatin, respectively, to prepare a material for wound dressing. They produced a gelatin/PU blended nanofiber scaffold and examined these scaffold by contact angle, water uptake, mechanical property, recovery, and degradation tests, and cellular response. They observed that, with the decreasing amount of gelatin in the blended solution, the contact angle increased, in the meanwhile water uptake of the scaffold decreased. The mechanical tests showed that, the blended nanofibrous scaffolds had an elastic character, as expected, and the elasticity of the scaffold increased with the increasing amount of PU. Beside, gelatin

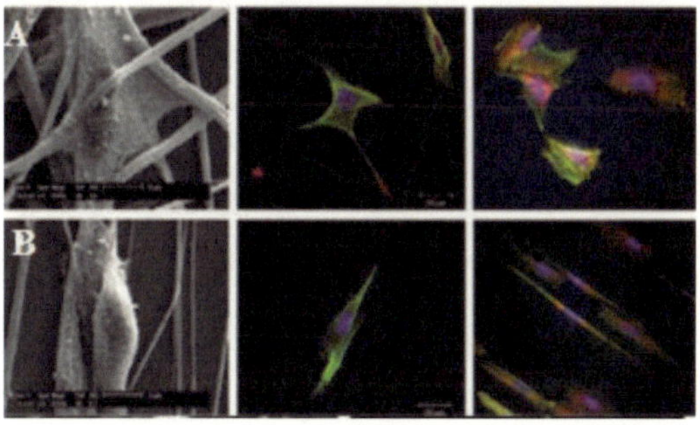

Figure 12. SEM images of (a) cells on randomly oriented nanofibers, (b) cells on aligned nanofibers and their fluorescence images [121].

amount increased the cell proliferation with the same amount of culture time. These results indicated that produced gelatin/PU blended nanofiber scaffold has a potential for wound dressing applications.

Lee et al. produced nanofiber-based artificial renal microfluidic chip by electrospinning method [122]. Authors have developed polyethersulfone and PU-based nanofibers and combined these webs with the poly(dimethylsiloxane)-based microfluidic platform top to create a chip-based portable hemodialysis system. They measured the filtration capability of this dialyzing chip and found that during the filtration and the transportation of the blood cells, they were not mechanically affected.

5.2. Other applications

In addition to biomedical applications, other application fields such as filtration, sensors, nanoweb lamination based on electrospun PU polymer nanofibers have been steadily extended in recent years. One main interest of electrospun polymer nanofiber non-woven mesh is for filtration application and it is best represented by relevant academic researches, in which many applications are in the field of filtration systems.

Filtration efficiency is closely associated with the fiber fineness and it is one of the most important concerns for the filter performance [10]. In general, due to the nanodimensions of the nanofiber, nanofiber mats have very high surface area to volume ratio and which results in high surface cohesion. Hence, tiny particles of the order of <0.5 mm can be easily trapped in the electrospun nanofibrous-structured filters. Thus, it is possible to improve the filtration efficiency with nanofibrous materials. In this manner, PU-based electrospun nanofibers could also be used as a filtration material because they are resistant to microorganisms and abrasion, and also have high hydrolytic stability [40].

Sambaer et al. [45], synthesized a PU based on 4,4'methylenebis(phenylisocyanate) (MDI), poly(3-methyl-1,5-pentanediol)-alt-(adipic, isophtalic acid) (PAIM) and 1,4 butanediol (BD) in molar ratio 9:1:8 at 90°C for 5 h. They used needless electrospinning apparatus and a supporting polyester fabric for collecting nanofibers with a speed of 0.16 m/min. Nanofibers were collected on the fabric with square ordering of electro conductive 9601 of fiber resistant, 95 g/m^2 of area mass, electro conductive 5 mm of thread distance. They evaluated the filtration efficiency of the nanofibers by experimental particle penetration efficiency according to EN 779 standard at the constant air flow rate 5.7 cm/s with aerosol particles. They also created 3D structure model from SEM image of the filter and compared this 3D structure model representing real filter structure with the corresponding experimental data. They obtained good agreement between both datasets.

In addition to fulfill the more traditional purpose in filtration, the nanofiber membranes fabricated from some specific polymers or coated with some selective agents can also be used as, for example, molecular filters or affinity membrane applications. Such filters can be applied to the detection and filtration of chemical and biological weapon agents [10].

Air filters separate particles mainly by the physical entrapment but also electrokinetic capture plays an important role in the air filter [46]. Through the filtration process particles collide with the filter medium. The filter medium that has an electrical charge and has attractive forces on the surface capture the particles. Due to attractive forces between charges or induced forces, particle deposition can occur on the surface. PU cationomers (PUCs) containing different amounts of quaternary ammonium groups were synthesized and successfully electrospun into non-woven nanofiber mats for use in antimicrobial nanofilter applications in the study of Jeong et al. [46]. The PUCs showed antimicrobial activities against *S. aureus* and *E. coli*. Due to the increased charge density of the PUC solutions, the average fiber diameters decreased with increasing quaternary ammonium group content. The PUC nanofibers showed adhesion between nanofibers with various bonding sites, yielding mats with a film-like character and structural integrity. The authors specified that, particle deposition could occur due to attractive forces between charges or induced forces. Therefore, they expected that the developed PUC nanofiber mats can exhibit a better performance as air nanofilters due to their surface electrical charges.

Ouyang et al. [123] reported the preparation of PU filled with carbon nanotubes and Ag nanoparticles (PU-MWCNT-AgNP) and the subsequent fabrication of a novel non-enzymatic amperometric biosensor for analytical determination of hydrogen peroxide. They conducted cyclic voltammetry experiments to indicate PU-MWCNT-AgNP nanofiber-modified electrodes have high electrocatalytic activity on hydrogen peroxide. The authors also carried out chronoamperometry measurements to illustrate developed electrospun sensor has high sensitivity for detecting hydrogen peroxide. Their study confirms that there is a remarkable synergistic effect of MWCNTs and AgNPs on the significant improvement of the conductivity of electrospun nanofibers. MWCNTs and AgNPs filling also affect the electrocatalytic activity, and the sensitivity of the fabricated non-enzymatic sensor. Their results indicated that the created biosensor for detecting hydrogen peroxide has a sensitivity of 160.6 $\mu A\ mM^{-1}\ cm^{-2}$, a wide linear range from 0.5 to 30 mM and a detection limit of 18.6 μM (S/N ¼ 3) and they claimed that PU–MWCNT–AgNP nanofibers have wide potential applications in bio-analysis and detection.

In most electrospinning setup, a conductive material is used as a collector, then produced nanofiber mat is removed from this material. But for some applications it is possible to use a nowoven or casual fabric as supporting material. US20100304108 describes a stretchable, non-woven nanofiber fabric. It allows vapor transport, capable of conforming to body parts but is impermeable to water. This breathable fabric could be useful in high-performance apparels and personal care products [124]. Inventors combined the fabric with different substrates to form a laminate (**Figure 13a**). Illustration of a composite breathable fabrics with electrospun membrane is given in **Figure 13b** [125]. Ahn et al. [24], electrospun PU nanofibers on to a water repellent nylon fabric and compared its waterproof and breathable properties with sole nylon fabric and polytetrafluoroethylene laminated nylon fabric. The aim of the authors is to develop an outdoor clothing with increased performance. A nanoweb laminate was prepared by laminating a PU electrospun nanoweb to the face fabric followed by heat treatment at 160°C in a tenter. The authors evaluated the water resistance and water vapor transmittance of the fabrics were under simulated microclimate. They examined the clothing microclimate and subjective sensations under normal and rainy atmospheric conditions. Their results indicated that the nanoweb laminate compared to the polytetrafluoroethylene laminate, had a higher water

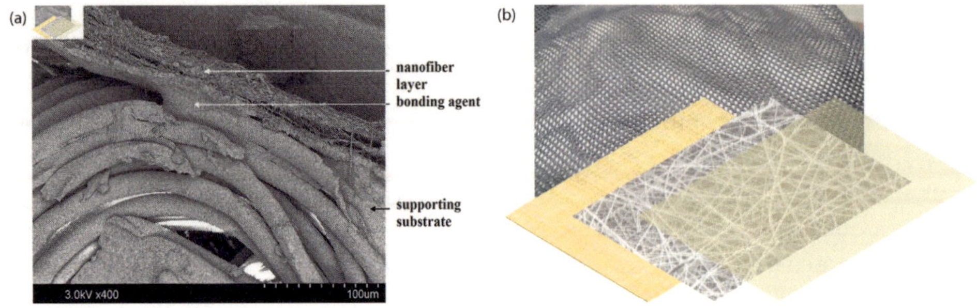

Figure 13. (a) A non-woven layer laminated fabric (US20100304108). (b) Illustration of a composite breathable fabrics with electrospun membrane [125].

vapor transmission rate but lower water resistance. They conducted wearing tests in a normal, warm environment to simulate exercising or sweating, to reveal the PU nanoweb-laminated clothing provided a more comfortable clothing microclimate than polytetrafluoroethylene-laminated clothing. In the rainy test conditions, they did not observe any difference between the polytetrafluoroethylene and the nanoweb-laminated clothing.

6. Conclusions

The main advantages of electrospinning process are the production of nanometer scale fibers with large surface areas, good mechanical properties, easy functionalization of these fibers for various purposes such as protective clothing, drug release material, or an air filter. These advantages provide many opportunities for their use in wide range of different applications [126]. PU is often chosen as a material for composing a nanoweb due to its chemical stability, mass transport, good mechanical properties, and also excellent nanofiber forming characteristic. Main reason for chosing PUs in electrospinning is that, the properties of PUs can vary in a wide range, so it is possible to adopt PUs to many applications by varying the structure of PUs. Electrospun PU nanofiber mats exhibiting good mechanical properties may have a wide variety of potential applications in high-performance air filters, protective textiles, wound dressing materials, sensors, in biomedical applications, drug delivery, as well as in wound dressing studies because of its good barrier properties and oxygen permeability. For these applications, electrospinning assembly can be varied in different ways for combining materials properties with different morphological structures. PUs can also be mixed either with a natural polymer or with a synthetic polymer for special applications such as collagen, dextran, and hydroxypropyl cellulose (HPC). Beside, from an industrial point of view, an enhancement in both mechanical and physical properties of the electrospun fiber mats is very important and it is possible to mix the PUs with SWCNTs or MWCNTs to overcome this problem.

Although there is ever increasing literature on use of PU nanofibers for various applications but still the field is in its infancy. For example, there are also lots of questions for biomedical applications, like interaction of scaffolds with biological systems, toxicity, in vivo studies etc.

http://dx.doi.org/10.5772/intechopen.69937

There is no doubt that PU electrospun materials are going to take major place in future for above mentioned applications. Thus, all these applications have to be thoroughly investigated before the technology can be used for any real practical application.

Author details

Cigdem Akduman[1] and Emriye Perrin Akçakoca Kumbasar[2]*

*Address all correspondence to: perrin.akcakoca@ege.edu.tr

1 Department of Textile Technology, Denizli Vocational School of Technical Sciences, Pamukkale University, Denizli, Turkey

2 Department of Textile Engineering, Faculty of Engineering, Ege University, Izmir, Turkey

References

[1] Wendorff JH, Agarwal S, Greiner A, editors. Electrospinning: Materials, Processing, and Applications. Weibheim, Germany: John Wiley & Sons; 2012. p. 241. ISBN: 978-3-527-64773-6

[2] Lukáš D, Sarkar A, Martinová L, Vodsed'álková K, Lubasová D, Chaloupek J, Pokorný P, Mikeš P, Chvojka J and Komárek M. Physical principles of electrospinning (Electrospinning as a nano-scale technology of the twenty-first century). Textile Progress. 2009;**41**(2):59-140. DOI: 10.1080/00405160902904641

[3] Cooley JF. Apparatus for Electrically Dispersing Fluids. U.S. Patent No. 692,631. Washington, DC: U.S. Patent and Trademark Office; 1902

[4] Anton process and apparatus for preparing artificial threads. U.S. Patent No. 1,975,504. Washington, DC: U.S. Patent and Trademark Office; 1934

[5] Doshi J and Reneker DH. Electrospinning process and application of electrospun fibers. Journal of Electrostatic. 1995;**35**:151-160. DOI: 10.1016/0304-3886(95)00041-8

[6] Srinivasan G and Reneker DH. Structure and morphology of small diameter electrospun aramid fibers. Polymer International. 1995;**36**(2):195-201. DOI: 10.1002/pi.1995.210360210

[7] Reneker DH and Chun I. Nanometre diameter fibres of polymer, produced by electrospinning. Nanotechnology. 1996;**7**(3):216-223

[8] Andrady AL. Science and Technology of Polymer Nanofibers. New Jersey: John Wiley & Sons; 2008. p. 403

[9] Ramakrishna S, Fujihara K, Teo W, Lim T, Ma Z. An Introduction to Electrospinning and Nanofibers. Singapore: World Scientific Publishing Co. Pte. Ltd.; 2005; 126-127. pp. 285-291

[10] Huang ZM, Zhang YZ, Kotaki M, Ramakrishna S. A review on polymer nanofibers by electrospinning and their applications in nanocomposites. Composites Science and Technology. 2003;**63**(15):2223-2253. DOI: 10.1016/S0266-3538(03)00178-7

[11] Wang B, Wang Y, Yin T and Yu Q. Applications of electrospinning technique in drug delivery. Chemical Engineering Communications. 2010;**197**(10):1315-1338. DOI: 10.1080/00986441003625997

[12] Sill TJ and Von Recum HA. Electrospinning: Applications in drug delivery and tissue engineering. Biomaterials. 2008;**29**(13):1989-2006. DOI: 10.1016/j.biomaterials.2008.01.011

[13] Teo WE and Ramakrishna S. A review on electrospinning design and nanofibre assemblies. Nanotechnology. 2006;**17**(14):R89. DOI: 10.1088/0957-4484/17/14/R01

[14] Subbiah T, Bhat GS, Tock RW, Parameswaran S, Ramkumar SS. Electrospinning of nanofibers. Journal of Applied Polymer Science. 2005;**96**(2):557-569. DOI: 10.1002/app.21481

[15] Taepaiboon P, Rungsardthong U, Supaphol P. Effect of cross-linking on properties and release characteristics of sodium salicylate-loaded electrospun poly (vinyl alcohol) fibre mats. Nanotechnolog. 2007;**18**(17):175102. DOI: 10.1088/0957-4484/18/17/175102

[16] Ngawhirunpat T, Opanasopit P, Rojanarata T, Akkaramongkolporn P, Ruktanonchai U, Supaphol P. Development of meloxicam-loaded electrospun polyvinyl alcohol mats as a transdermal therapeutic agent. Pharmaceutical Development and Technology. 2009;**14**(1):73-82. DOI: 10.1080/10837450802409420

[17] Wang L and Ryan AJ. Introduction to electrospinning. In: Electrospinning for Tissue Cambridge: Woodhead Publishing; 2011. pp. 3-33. 409p

[18] Greiner A, Wendorff JH. Electrospinning: A fascinating method for the preparation of ultrathin fibers. Angewandte Chemie International Edition. 2007;**46**(30):5670-5703. DOI: 10.1002/anie.200604646

[19] Sen R, Zhao B, Perea D, Itkis ME, Hu H, Love J, Bekyarova E, Haddon RC. Preparation of single-walled carbon nanotube reinforced polystyrene and polyurethane nanofibers and membranes by electrospinning. Nano Letters. 2004;**4**(3):459-464. DOI: 10.1021/nl035135s

[20] Demir MM, Yilgor I, Yilgor EEA and Erman B. Electrospinning of polyurethane fibers. Polymer. 2002;**43**(11):3303-3309. DOI: 10.1016/S0032-3861(02)00136-2

[21] Khil MS, Cha DI, Kim HY, Kim IS, Bhattarai N. Electrospun nanofibrous polyurethane membrane as wound dressing. Journal of Biomedical Materials Research Part B: Applied Biomaterial. 2003;**67**(2):675-679. DOI: 10.1002/jbm.b.10058

[22] Barick AK and Tripathy DK. Thermal and dynamic mechanical characterization of thermoplastic polyurethane/organoclay nanocomposites prepared by melt compounding. Materials Science and Engineering: A. 2010;**527**(3):812-823. DOI: 10.1016/j.msea.2009.10.063

[23] Huynh TTN, Padois K, Sonvico F, Rossi A, Zani F, Pirot F, Doury J and Falson F. Characterization of a polyurethane-based controlled release system for local delivery

of chlorhexidine diacetate. European Journal of Pharmaceutics and Biopharmaceutics. 2010;**74**(2):255-264. DOI: 10.1016/j.ejpb.2009.11.002

[24] Ahn HW, Park CH, Chung SE. Waterproof and breathable properties of nanoweb applied clothing. Textile Research Journal. 2011;**81**(14):1438-1447. DOI: 10.1177/0040517510392462

[25] Martin DJ, Osman AF, Andriani Y and Edwards GA. Thermoplastic polyurethane (TPU)-based polymer nanocomposites. In: Advances in Polymer Nanocomposites: Types and Applications. Cambridge: Woodhead Publishing Limited; 2012. p. 651

[26] Mi H-Y, Jing X, Jacques BR, Turng L-S, Peng X-F. Characterization and properties of electrospun thermoplastic polyurethane blend fibers: Effect of solution rheological properties on fiber formation. Journal of Materials Research. 2013;**28**(17):2339-2350. DOI: 10.1557/jmr.2013.115

[27] Yanilmaz M, Kalaoglu F, Karakas H. Investigation on the effect of process variables on polyurethane nanofibre diameter using a factorial design. FIBRES & TEXTILES in Eastern Europe. 2013;**21**, 2(98):19-21

[28] Kumbasar EPA, Akduman Ç, Çay A. Effects of β-cyclodextrin on selected properties of electrospun thermoplastic polyurethane nanofibres. Carbohydrate Polymers. 2014;**104**:42-49. DOI: 10.1016/j.carbpol.2013.12.065

[29] Mondal S. Influence of solvents properties on morphology of electrospun polyurethane nanofiber mats. Polymers Advanced Technologies. 2014;**25**:179-183. DOI: 10.1002/pat.3220

[30] Sperling LH. Introduction to Physical Polymer Science. New Jersey. John Wiley & Sons; 2006. p. 845

[31] https://en.wikipedia.org/wiki/Polyurethane [Accessed: 2017-01-25]

[32] Petrovic ZS. Polyurethanes. In: Kricheldorf HR, Nuyken O, Swift G, editors. Handbook of Polymer Synthesis. New York: Marcel Dekker; 2005. p. 965

[33] Ulrich H. Polyurethanes. In: Mark HF, editor. Encyclopedia of Polymer Science and Technology. 3rd ed. Vol. 4. Hoboken: Wiley; 2003

[34] Kultys A, Rogulska M, Głuchowska H. The effect of soft-segment structure on the properties of novel thermoplastic polyurethane elastomers based on an unconventional chain extender. Polymer International. 2011;**60**(4):652-659. DOI: 10.1002/pi.2998

[35] http://www.huntsman.com/polyurethanes/Media%20Library/global/files/guide_tpu.pdf [Accessed: 2017-01-25]

[36] Yao C, Li X, Neoh KG, Shi Z, Kang ET. Surface modification and antibacterial activity of electrospun polyurethane fibrous membranes with quaternary ammonium moieties. Journal of Membrane Science. 2008;**320**(1):259-267. DOI: 10.1016/j.memsci.2008.04.012

[37] Zdrahala RJ and Zdrahala IJ. Biomedical applications of polyurethanes: A review of past promises, present realities, and a vibrant future. Journal of Biomaterials Applications. 1999;**14**(1):67-90. DOI: 10.1177/088532829901400104

[38] Zhuo H, Hu J, Chen S, Yeung L. Preparation of polyurethane nanofibers by electrospinning. Journal of Applied Polymer Science. 2008;**109**(1):406-411. DOI: 10.1002/app.28067

[39] Kim SE, Heo DN, Lee JB, Kim JR, Park SH, Jeon SH and Kwon IK. Electrospun gelatin/polyurethane blended nanofibers for wound healing. Biomedical Materials. 2009;**4**(4):044106. DOI: 10.1088/1748-6041/4/4/044106

[40] Kimmer D, Slobodian P, Petráš D, Zatloukal M, Olejník R, Sáha P. Polyurethane/multiwalled carbon nanotube nanowebs prepared by an electrospinning process. Journal of Applied Polymer Science. 2009;**111**(6):2711-2714. DOI: 10.1002/app.29238

[41] Vlad S, Ciobanu C, Macocinschi D, Filip D, Butnaru M, Gradinaru LM, Nistor A. Polyurethane nanofibers by electrospinning for biomedical applications. In: CAS 2010 Proceedings (International Semiconductor Conference) IEEE; October 2010. Vol. 2. pp. 353-356

[42] Karchin A, Simonovsky FI, Ratner BD, Sanders JE. Melt electrospinning of biodegradable polyurethane scaffolds. Acta Biomaterialia. 2011;**7**(9):3277-3284. DOI: 10.1016/j.actbio.2011.05.017

[43] Unnithan AR, Barakat NA, Pichiah PT, Gnanasekaran G, Nirmala R, Cha YS, Jung CH, El-Newehy M and Kim HY. Wound-dressing materials with antibacterial activity from electrospun polyurethane–dextran nanofiber mats containing ciprofloxacin HCl. Carbohydrate Polymers. 2012;**90**(4):1786-1793. DOI: 10.1016/j.carbpol.2012.07.071

[44] Hong JH, Jeong EH, Lee HS, Baik DH, Seo SW, Youk JH. Electrospinning of polyurethane/organically modified montmorillonite nanocomposites. Journal of Polymer Science Part B: Polymer Physics. 2005;**43**(22):3171-3177. DOI: 10.1002/polb.20610

[45] Sambaer W, Zatloukal M, Kimmer D. 3D modeling of filtration process via polyurethane nanofiber based nonwoven filters prepared by electrospinning process. Chemical Engineering Science. 2011;**66**(4):613-623. DOI: 10.1016/j.ces.2010.10.035

[46] Jeong EH, Yang J, Youk JH. Preparation of polyurethane cationomer nanofiber mats for use in antimicrobial nanofilter applications. Materials Letters. 2007;**61**(18):3991-3994. DOI: 10.1016/j.matlet.2007.01.003

[47] Wannatong L, Sirivat A and Supaphol P. Effects of solvents on electrospun polymeric fibers: Preliminary study on polystyrene. Polymer International. 2004;**53**:1851-1859. DOI: 10.1002/pi.1599

[48] Yanılmaz M, Kalaoğlu F, Karakaş H. Study on optimising the morphology of electrospun polyurethane nanofibers. Tekstil ve Konfeksiyon. 2012;**22**(3):212-217

[49] Dasdemir M, Topalbekiroglu M, Demir A. Electrospinning of thermoplastic polyurethane microfibers and nanofibers from polymer solution and melt. Journal of Applied Polymer Science. 2013;**127**:1901-1908. DOI: 10.1002/app.37503

[50] http://www.rowak.ch/en/glossary/Polyurethane.php [Accessed: 2017-01-25]

[51] Oprea S. Effect of solvent interactions on the properties of polyurethane films. High Performance Polymers. 2005;**17**(2):163-173. DOI: 10.1177/0954008305045688

[52] Çay A, Akçakoca Kumbasar EP, Akduman Ç. Effects of solvent mixtures on the morphology of electropsun thermoplastic polyurethane nanofibres. Tekstil ve Konfeksiyon. 2015;**25**(1):38-46

[53] http://cccbdb.nist.gov/diplistx.asp [Accessed: 2017-01-25]

[54] http://www.stenutz.eu/chem/solv6.php?name=N,N-dimethylacetamide [Accessed: 2017-01-25]

[55] http://cool.conservation-us.org/coolaic/sg/bpg/annual/v03/bp03-04.html [Accessed: 2017-01-25]

[56] http://clas.sa.ucsb.edu/staff/Resource%20Folder/Chem109ABC/Substitution%20Reactions/Dielectric%20Constants%20of%20Common%20Solvents.pdf [Accessed: 2017-01-25]

[57] Kidoaki S, Kwon IK, & Matsuda T. Structural features and mechanical properties of in situ–bonded meshes of segmented polyurethane electrospun from mixed solvents. Journal of Biomedical Materials Research Part B: Applied Biomaterials. 2006;**76**(1):219-229. DOI: 10.1002/jbm.b.30336

[58] Akçakoca Kumbasar EP, Akduman Ç, & Çay A. Electrospinning parameters for thermoplastic polyurethane (TPU) nanofibers. In: Proceedings of the 13th AUTEX World Textile Conference; 22-24 May 2013; Dresden, Germany

[59] Andrews KD, Hunt JA, & Black RA. Technology of electrostatic spinning for the production of polyurethane tissue engineering scaffolds. Polymer International. 2008;**57**(2):203-210. DOI: 10.1002/pi.2317

[60] Cengiz F and Jirsak O. The effect of salt on the roller electrospinning of polyurethane nanofibers. Fibers and Polymers. 2009;**10**(2):177-184. DOI: 10.1007/s12221-009-0177-7

[61] Cengiz-Çallıoğlu F, Jirsak O and Dayık M. The influence of non-solvent addition on the independent and dependent parameters in roller electrospinning of polyurethane. Journal of Nanoscience and Nanotechnology. 2013;**13**(7):4727-4735. DOI: 10.1166/jnn.2013.7190

[62] Yalcinkaya B, Callioglu FC and Yener F. Measurement and analysis of jet current and jet life in roller electrospinning of polyurethane. Textile Research Journal. 2014;**84**(16):1720-1728. DOI: 10.1177/0040517514528563

[63] Takasaki M, Sugihara K, Ohkoshi Y, Fujii T, Shimizu H, & Saito M. Thermoplastic polyurethane ultrafine fiber web fabricated by laser electrospinning. Sen'i Gakkaishi. 2010;**66**(7):168-173

[64] Pedicini A and Farris RJ. Mechanical behavior of electrospun polyurethane. Polymer. 2003;**44**(22):6857-6862. DOI: 10.1016/j.polymer.2003.08.040

[65] Theron JP, Knoetze JH, Sanderson RD, Hunter R, Mequanint K, Franz T and Bezuidenhout D. Modification, crosslinking and reactive electrospinning of a thermoplastic medical polyurethane for vascular graft applications. Acta Biomaterialia. 2010;**6**(7):2434-2447. DOI: 10.1016/j.actbio.2010.01.013

[66] Teo WE, Kotaki M, Mo XM and Ramakrishna S. Porous tubular structures with controlled fibre orientation using a modified electrospinning method. Nanotechnology. 2005;**16**(6):918. DOI: 10.1088/0957-4484/16/6/049

[67] Kim KW, Lee KH, Khil MS, Ho YS and Kim HY. The effect of molecular weight and the linear velocity of drum surface on the properties of electrospun poly (ethylene terephthalate) nonwovens. Fibers and Polymers. 2004;**5**(2):122-127

[68] Matthews JA, Wnek GE, Simpson DG, & Bowlin GL. Electrospinning of collagen nanofibers. Biomacromolecules. 2002;**3**(2):232-238. DOI: 10.1021/bm015533u

[69] Bhattarai N, Edmondson D, Veiseh O, Matsen FA and Zhang M. Electrospun chitosan-based nanofibers and their cellular compatibility. Biomaterials. 2005;**26**(31):6176-6184. DOI: 10.1016/j.biomaterials.2005.03.027

[70] Li D, Wang Y and Xia Y. Electrospinning of polymeric and ceramic nanofibers as uniaxially aligned arrays. Nano Letters. 2003;**3**(8):1167-1171. DOI: 10.1021/nl0344256

[71] Katta P, Alessandro M, Ramsier RD and Chase GG. Continuous electrospinning of aligned polymer nanofibers onto a wire drum collector. Nano Letters. 2004;**4**(11):2215-2218. DOI: 10.1021/nl0486158

[72] Theron A, Zussman E and Yarin AL. Electrostatic field-assisted alignment of electrospun nanofibres. Nanotechnology. 2001;**12**(3):384. DOI: 10.1088/0957-4484/12/3/329

[73] Zussman E, Theron A and Yarin AL. Formation of nanofiber crossbars in electrospinning. Applied Physics Letters. 2003;**82**(6):973-975. DOI: 10.1063/1.1544060#

[74] Xu CY, Inai R, Kotaki M and Ramakrishna S. Aligned biodegradable nanofibrous structure: A potential scaffold for blood vessel engineering. Biomaterials. 2004;**25**(5):877-886. DOI: 10.1016/S0142-9612(03)00593-3

[75] Li D, Wang Y and Xia Y. Electrospinning nanofibers as uniaxially aligned arrays and layer-by-layer stacked films. Advanced Materials. 2004;**16**(4):361-366. DOI: 10.1002/adma.200306226

[76] Teo WE and Ramakrishna S. Electrospun fibre bundle made of aligned nanofibres over two fixed points. Nanotechnology. 2005;**16**(9):1878. DOI: 10.1088/0957-4484/16/9/077

[77] Dalton PD, Klee D and Möller M. Electrospinning with dual collection rings. Polymer. 2005;**46**(3):611-614. DOI: 10.1016/j.polymer.2004.11.075

[78] https://www.uab.edu/cas/physics/graduate/48-research/116-research-vohra-02 [Accessed: 2017-01-25]

[79] Banuskeviciute A, Adomaviciute E, Milasius R and Stanys S. Formation of thermoplastic polyurethane (TPU) nano/micro fibers by electrospinning process using electrode with tines. Materials Science. 2011;**17**(3):287-292

[80] Thandavamoorthy S, Gopinath N and Ramkumar SS. Self-assembled honeycomb polyurethane nanofibers. Journal of Applied Polymer Science. 2006;**101**(5):3121-3124. DOI: 10.1002/app.24333

[81] http://www.nanowerk.com/news/newsid=777.php [Accessed: 2017-01-25]

[82] Chen JP, Chang GY, & Chen JK. Electrospun collagen/chitosan nanofibrous membrane as wound dressing. Colloids and Surfaces A: Physicochemical and Engineering Aspects. 2008;**313**:183-188. DOI: 10.1016/j.colsurfa.2007.04.129

[83] Buruaga L, Sardon H, Irusta L, González A, Fernández-Berridi MJ and Iruin JJ. Electrospinning of waterborne polyurethanes. Journal of Applied Polymer Science. 2010;**115**(2):1176-1179. DOI: 10.1002/app.31219

[84] Lee KH, Kim HY, Ryu YJ, Kim KW, Choi SW. Mechanical behavior of electrospun fiber mats of poly (vinyl chloride)/polyurethane polyblends. Journal of Polymer Science Part B: Polymer Physics. 2003;**41**(11):1256-1262. DOI: 10.1002/polb.10482

[85] Akçakoca Kumbasar EP, Akduman Ç and Çay A. Effects of β-cyclodextrin on selected properties of electrospun thermoplastic polyurethane nanofibres. Carbohydrate Polymers. 2014;**104**:42-49. DOI: 10.1016/j.carbpol.2013.12.065

[86] Hu W and Yu X. Thermal and mechanical properties of bio-based PCMs encapsulated with nanofibrous structure. Renewable Energy. 2014;**62**:454-458. DOI: 10.1016/j.renene.2013.07.047

[87] Unnithan AR, Gnanasekaran G, Sathishkumar Y, Lee YS and Kim CS. Electrospun antibacterial polyurethane–cellulose acetate–zein composite mats for wound dressing. Carbohydrate Polymers. 2014;**102**:884-892. DOI: 10.1016/j.carbpol.2013.10.070

[88] Chen R, Huang C, Ke Q, He C, Wang H and Mo X. Preparation and characterization of coaxial electrospun thermoplastic polyurethane/collagen compound nanofibers for tissue engineering applications. Colloids and Surfaces B: Biointerfaces. 2010;**79**(2):315-325. DOI: 10.1016/j.colsurfb.2010.03.043

[89] http://electrospintech.com/core-shellintro.html#.WKbnHYVOL4c [Accessed: 2017-01-25]

[90] Huang C, Chen R, Ke Q, Morsi Y, Zhang K and Mo X. Electrospun collagen–chitosan–TPU nanofibrous scaffolds for tissue engineered tubular grafts. Colloids and Surfaces B: Biointerfaces. 2011;**82**(2):307-315. DOI: 10.1016/j.colsurfb.2010.09.002

[91] Chen JP and Chiang Y. Bioactive electrospun silver nanoparticles-containing polyurethane nanofibers as wound dressings. Journal of Nanoscience and Nanotechnology. 2010;**10**(11):7560-7564. DOI: 10.1166/jnn.2010.2829

[92] Lee K, Lee B, Kim C, Kim H, Kim K and Nah C. Stress-strain behavior of the electrospun thermoplastic polyurethane elastomer fiber mats. Macromolecular Research. 2005;**13**(5):441-445. DOI: 10.1007/BF03218478

[93] Cha DI, Kim HY, Lee KH, Jung YC, Cho JW and Chun BC. Electrospun nonwovens of shape-memory polyurethane block copolymers. Journal of Applied Polymer Science. 2005;**96**(2):460-465. DOI: 10.1002/app.21467

[94] Polymer/Carbon Nanotube Nanocomposites [Accessed: 2017-01-25]

[95] Liu L and Pan S. Comparative study of the comprehensive properties of multiwalled nanotubes/polyurethane nanofibrous membranes with and without thermal treatment. Journal of Nanomaterials. 2012;**2012**:9. DOI: 10.1155/2012/610781

[96] Verreck G, Chun I, Rosenblatt J, Peeters J, Van Dijck A, Mensch J, Noppe M and Brewster ME. Incorporation of drugs in an amorphous state into electrospun nanofibers composed of a water-insoluble, nonbiodegradable polymer. Journal of Controlled Release. 2003;**92**(3):349-360. DOI: 10.1016/S0168-3659(03)00342-0

[97] Tungprapa S, Jangchud I and Supaphol P. Release characteristics of four model drugs from drug-loaded electrospun cellulose acetate fiber mats. Polymer. 2007;**48**(17):5030-5041. DOI: 10.1016/j.polymer.2007.06.0

[98] Taepaiboon P, Rungsardthong U and Supaphol P. Drug-loaded electrospun mats of poly (vinyl alcohol) fibres and their release characteristics of four model drugs. Nanotechnology. 2006;**17**(9):2317. DOI: 10.1088/0957-4484/17/9/041

[99] Akduman C, Özgüney I and Akçakoca Kumbasar EP. Preparation and characterization of naproxen-loaded electrospun thermoplastic polyurethane nanofibers as a drug delivery system. Materials Science and Engineering: C. 2016;**64**:383-390. DOI: 10.1016/j.msec.2016.04.005

[100] Kenawy ER, Abdel-Hay FI, El-Newehy MH and Wnek GE. Controlled release of ketoprofen from electrospun poly (vinyl alcohol) nanofibers. Materials Science and Engineering: A. 2007;**459**(1):390-396. DOI: 10.1016/j.msea.2007.01.039

[101] Kenawy ER, Abdel-Hay FI, El-Newehy MH and Wnek GE. Processing of polymer nanofibers through electrospinning as drug delivery systems. Materials Chemistry and Physics. 2009;**113**(1):296-302. DOI: 10.1016/j.matchemphys.2008.07.081

[102] Khadka DB and Haynie DT. Protein-and peptide-based electrospun nanofibers in medical biomaterials. Nanomedicine: Nanotechnology, Biology and Medicine. 2012;**8**(8):1242-1262. DOI: 10.1016/j.nano.2012.02.013

[103] Tian M, Gao Y, Liu Y, Liao Y, Xu R, Hedin NE and Fong H. Bis-GMA/TEGDMA dental composites reinforced with electrospun nylon 6 nanocomposite nanofibers containing highly aligned fibrillar silicate single crystals. Polymer. 2007;**48**(9):2720-2728. DOI: 10.1016/j.polymer.2007.03.032

[104] Liao IC, Chew SY and Leong KW. Aligned core–shell nanofibers delivering bioactive proteins. Nanomedicine. 2006;**1**(4):465-471. DOI: 10.2217/17435889.1.4.465

[105] Jiang H, Wag L and Zhu K. Coaxial electrospinning for encapsulation and controlled release of fragile water-soluble bioactive agents. Journal of Controlled Release. 2014;**193**:296-303. DOI: 10.1016/j.jconrel.2014.04.025

[106] Nisbet DR, Forsythe JS, Shen W, Finkelstein DI and Horne MK. Review paper: A review of the cellular response on electrospun nanofibers for tissue engineering. Journal of Biomaterials Applications. 2008;**24**:7-29. DOI: 10.1177/0885328208099086

[107] Li D, Frey MW and Baeumner AJ. Electrospun polylactic acid nanofiber membranes as substrates for biosensor assemblies. Journal of Membrane Science. 2006;**279**(1):354-363. DOI: 10.1016/j.memsci.2005.12.036

[108] Ren G, Xu X, Liu Q, Cheng J, Yuan X, Wu L and Wan Y. Electrospun poly (vinyl alcohol)/glucose oxidase biocomposite membranes for biosensor applications. Reactive and Functional Polymers. 2006;**66**(12):1559-1564. DOI: 10.1016/j.reactfunctpolym.2006.05.005

[109] Yang Q, Wu J, Li JJ, Hu MX and Xu ZK. Nanofibrous sugar sticks electrospun from glycopolymers for protein separation via molecular recognition. Macromolecular Rapid Communications. 2006;**27**(22):1942-1948. DOI: 10.1002/marc.200600470

[110] Qin XH and Wang SY. Filtration properties of electrospinning nanofibers. Journal of Applied Polymer Science. 2006;**102**(2):1285-1290. DOI: 10.1002/app.24361

[111] Gopal R, Kaur S, Feng CY, Chan C, Ramakrishna S, Tabe S and Matsuura T. Electrospun nanofibrous polysulfone membranes as pre-filters: Particulate removal. Journal of Membrane Science. 2007;**289**(1):210-219. DOI: 10.1016/j.memsci.2006.11.056

[112] Gorji M, Jeddi A and Gharehaghaji AA. Fabrication and characterization of polyurethane electrospun nanofiber membranes for protective clothing applications. Journal of Applied Polymer Science. 2012;**125**(5):4135-4141. DOI: 10.1002/app.36611

[113] Lee S and Obendorf SK. Use of electrospun nanofiber web for protective textile materials as barriers to liquid penetration. Textile Research Journal. 2007;**77**(9):696-702. DOI: 10.1177/0040517507080284

[114] Schreuder-Gibson HL, Truong Q, Walker JE, Owens JR, Wander JD and Jones WE. Chemical and biological protection and detection in fabrics for protective clothing. MRS Bulletin. 2003;**28**(08):574-578

[115] Ma Z, Kotaki M, Inai R and Ramakrishna S. Potential of nanofiber matrix as tissue-engineering scaffolds. Tissue Engineering. 2005;**11**(1-2):101-109. DOI: 10.1089/ten.2005.11.101

[116] Amoroso NJ, D'Amore A, Hong Y, Wagner WR and Sacks MS. Elastomeric electrospun polyurethane scaffolds: The interrelationship between fabrication conditions, fiber topology, and mechanical properties. Advanced Materials. 2011;**23**(1):106-111. DOI: 10.1002/adma.201003210

[117] Carlberg B, Axell MZ, Nannmark U, Liu J and Kuhn HG. Electrospun polyurethane scaffolds for proliferation and neuronal differentiation of human embryonic stem cells. Biomedical Materials. 2009;**4**(4):045004. DOI: 10.1088/1748-6041/4/4/045004

[118] Grasl C, Bergmeister H, Stoiber M, Schima H and Weigel G. Electrospun polyurethane vascular grafts: In vitro mechanical behavior and endothelial adhesion molecule expression. Journal of Biomedical Materials Research Part A. 2010;**93**(2):716-723. DOI: 10.1002/jbm.a.32584

[119] Mitchell SB, Sanders JE. A unique device for controlled electrospinning. Journal of Biomedical Materials Research Part A. 2006;**78**(1):110-120. DOI: 10.1002/jbm.a.30673

[120] Sheikh FA, Barakat NA, Kanjwal MA, Chaudhari AA, Jung IH, Lee JH and Kim HY. Electrospun antimicrobial polyurethane nanofibers containing silver nanoparticles for biotechnological applications. Macromolecular Research. 2009;**17**(9):688-696. DOI: 10.1007/BF03218929

[121] Vajapeyajula S and Johnson R. Nanofiber Scaffold Technology Validated in a Historic Human Implant, Biowire Spring 2012. pp. 31-34, http://www.sigmaaldrich.com/technical-documents/articles/biowire/nanofiber-scaffold-technology.html [Accessed: 2017-01-25]

[122] Lee KH, Kim DJ, Min BG and Lee SH. Polymeric nanofiber web-based artificial renal microfluidic chip. Biomedical Microdevices. 2007;**9**(4):435-442. DOI: 10.1007/s10544-007-9047-5

[123] Ouyang Z, Li J, Wang J, Li Q, Ni T, Zhang X, Wei G. Fabrication, characterization and sensor application of electrospun polyurethane nanofibers filled with carbon nanotubes and silver nanoparticles. Journal of Materials Chemistry B. 2013;**1**(18):2415-2424. DOI: 10.1039/C3TB20316F

[124] Doshi JN, Vora MM, Ooten G. Stretchable nonwoven fabric, method of manufacturing, and products made thereof. U.S. Patent Application No. 12/474,586; 2010

[125] http://electrospintech.com/breathablefabric.html#.WK2K-YVOL4c [Accessed: 2017-01-25]

[126] Agarwal S, Wendorff JH, Greiner A. Use of electrospinning technique for biomedical applications. Polymer. 2008;**49**(26):5603-5621. DOI: 10.1016/j.polymer.2008.09.014

Chapter 3

Polyurethane: A Shape Memory Polymer (SMP)

Suman Thakur and Jinlian Hu

Additional information is available at the end of the chapter

http://dx.doi.org/10.5772/intechopen.69992

Abstract

Shape memory polymer (SMP) is a stimuli-responsive material with the ability to alter a programmed shape to its original shape upon triggering of an appropriate stimulus. For the past decades, SMP has dragged much interest in material field owing to its various and versatile applications. One archetypal SMP is polyurethane, which has a wide-ranging transition temperature for its shape recovery, retraction temperatures, inherent soft-hard segments, a high recoverable strain (up to 400%), high control of the softening, favorable and tunable physical properties, and so on. This chapter emphasizes on the raw materials required for the synthesis of shape memory polyurethane (SMPU), the principle of shape memory function, the design of protocol of SMPU, and their applications with future directions.

Keywords: polyurethane, shape memory, smart, stimulus sensitive

1. Introduction

Shape memory polymer (SMP) is a stimuli-responsive polymer which has an ability to fix a temporarily deformed shape and to return from the deformed shape to its original permanent shape while induced by an external stimulus such as thermal, humidity, pH, light, magnetic energy, electric field, and so on [1–5]. SMP possesses two phases viz. frozen phase or net hard phase and reversible or switching soft phase. In case of polyurethane (PU), they may have reversible amorphous or crystalline phase [6, 7]. PU has a unique structural characteristic because of the presence of inherent incompatibility due to microphase in-homogeneity in its chain molecules. It possesses a wide-ranging temperature for shape recovery, high recoverable strain (up to 400%), inherent soft-hard segments, high control on the softening and retraction temperatures with good biophysical properties [8–11]. Further, the properties including switching transition temperature like glass transition temperature (T_g) or melting temperature (T_m) can be tuned much easily by proper choice and using suitable composition of the components during polymerization process. A large number of versatile components with different structures and properties are available for this polymerization [12]. Thus, PU is versatile with respect to its structure and properties. Further, it is available in various forms, thermoplastic, foam, elastomer, and fiber [13, 14]. Thus, among the different types of SMP, PU is a good candidate as SMP for various advanced applications. In this chapter, we focus on the raw materials required for the synthesis of shape memory polyurethane (SMPU), the principle of shape memory function, the design of protocol of SMPU, the recent research progress of SMPU, and their potential applications.

2. Raw materials for shape memory polyurethane

PU is an important subclass of polymers and it contains multi-disperse blocks of soft and hard segments in an alternative fashion. The urethane linkage (—NHCOO—) is produced by a rearrangement reaction between a diisocyanate and a polyol. The common raw materials used for PU synthesis are categorized into macroglyol, diisocyanate, and chain extender. The brief descriptions of these components are presented in this section.

2.1. Diisocyanate

It is a fundamental precursor for the synthesis of PU. Although both types of diisocyanate viz. aromatic and aliphatic are used, aromatic diisocyanates are most commonly used for SMPU. The properties of SMPU greatly depend on the structure of diisocyanate component. Thus, appropriate selection of diisocyanate component envisages the properties of SMPU. The most commonly used diisocyanates are toluene diisocyanate (TDI), 4,4-diphenyl methane diisocyanate (MDI), polymeric methylene diphenyl diisocyanate (PMDI), 3,3-dimethyl diphenyl methane diisocyanate (DDI), naphthalene diisocyanate (NDI), hexamethylene diisocyanate (HDI), isophorone diisocyanate (IPDI), and lysine diisocyanate (LDI).

Among them, TDI and MDI are widely used in the preparation of SMPU [15, 16]. In most of the cases, TDI is used as a mixture of the 2,4- and 2,6-isomers in 80:20 mol ratio. Similarly, MDI also has three isomers namely 4,4-, 2,4-, and 2,2-diphenyl methane diisocyanates. However, 4,4-isomer is used in most of the SMPUs [15]. Though aromatic diisocyanates have higher reactivity than aliphatic ones and obtained SMPU exhibited good thermal and mechanical properties, the former SMPU also suffers from a few inadequacies such as lower oxidation and weaker ultraviolet stabilization [17]. Therefore, the use of aliphatic diisocyanates is required, where color and transparency are important for end application.

2.2. Macroglycol

In the synthesis of SMPU, the aliphatic diols or polyols with a molecular weight of 400–5000 g mol^{-1} are mainly used as macroglycols. The long-chain macroglycol containing low functionality provides elastomeric soft SMPU, whereas the short-chain macroglycol with high functionality forms rigid SMPU. Various types of macroglycol such as polyester, polyether, polycarbonate, hydrocarbon, and so on are used in the synthesis of SMPU. However, the polyester and polyether macroglycols are mostly used among them. Polyester macroglycol provides high flexibility, which is very useful in SMPU. However, the ester groups are easily hydrolyzed by alkali, thereby weakening the performance of SMPU in special biological application under alkaline condition [18]. A few of polyester polyol macroglycols such as polycaprolactone, polycarbonate, and so on demonstrate high hydrolytic stability in alkaline condition too. On the other hand, polyether macroglycol contributes some specific properties such as metal chelation, hydrophilicity, crystallinity, surface activity, and so on. In addition to that, different dendritic or hyperbranched polyols are also used to prepare SMPU. Among all, the crystalline polycaprolactone diol (PCL) and polyethylene glycol (PEG) are widely used for SMPU as macroglycol.

2.3. Chain extender

Several low-molecular-weight (generally below 500 g/mol) diols, diamines, and amino alcohols are used as chain extenders for the synthesis of SMPU [18–20]. The most commonly used chain extenders are ethylene glycol, 1,4-butanediol, 1,6-hexanediol, diethylene triamine, and diethanol amine. The chain extender is generally used to increase the chain length and the molecular weight of SMPU. The amine chain extender reacts rapidly and increases the crosslinking density of SMPU by bridging with biuret linkages in case of SMPU thermoset, whereas diol chain extenders sometimes required organometallic catalysts to obtain SMPU. On the other hand, multifunctional and highly branched chain extenders like trimethylol propane, glycerol, triglyceride ricinolate, hyperbranched polyol, and so on are used to prepare hyperbranched structure in SMPU.

2.4. Catalyst

The isocyanates especially the aromatic ones are highly reactive, so catalyst is generally not needed in such cases. However, catalyst is required for synthesis of aliphatic isocyanate containing SMPU and at low temperature. Also, the reaction rate rapidly decreases when the steric hindrance increases, mainly on the substitute in higher alcohols. In such cases, catalyst is required. These catalysts are mainly divided into two categories viz. amino (basic) and organometallic compounds. The commonly used amine catalysts are triethyl amine, triethylenediamine (TEDA), dimethylethanolamine (DMEA), dimethylcyclohexylamine (DMCHA), and so on. Similarly, organometallic catalysts are mercury, lead, tin, bismuth, and zinc-based organo-complexes. However, these catalysts are toxic and hence they have disposal problems which hinder their uses. Nowadays, bismuth and zinc carboxylates, alkyl tin carboxylates, oxides and mercaptide oxides such as dibutyl tin dilaurate (DBTDL), dioctyltin mercaptide, stannous octoate, dibutyltin oxide, and so on are used successfully [21].

3. Properties and characterization of shape memory polyurethane

Transition temperature (T_{trans}), shape fixity (R_f), shape recovery ratio (R_r), maximum recoverable strain (ε_{max}), and maximum recovery stress (σ_{max}) are the important parameters that are used to describe shape memory effects of a polymeric material [22].

Shape fixity: Shape fixity is the extent of a temporary shape being fixed for an SMP. It is also known as strain fixity or shape retention. The shape fixity is thus equal to the percentage of the ratio of fixed deformation to total deformation (Eq. (1))

$$R_f = \frac{\varepsilon}{\varepsilon_{load}} \times 100\% \quad (1)$$

Shape recovery: Shape recovery is defined as the ability of a polymeric material to memorize the original shape from a temporary deformed shape. Therefore, the shape recovery is the percentage of the ratio of deformation recovered by the specimen to the deformation taken place to the specimen (Eq. (2))

$$R_r = \frac{\varepsilon - \varepsilon_{recovery}}{\varepsilon} \times 100\% \quad (2)$$

Recovery rate: This parameter describes the speed, that is, the rate of recovery from a programmed shape to its original shape upon triggering of a proper stimulus. It can also be said as the speed of recovery process or shape recovery speed.

T_{trans} is usually equal to T_g for an amorphous SMPU or T_m for a crystalline SMPU. This is generally measured by standard thermal analysis techniques such as differential scanning calorimetry (DSC) or dynamic mechanical analysis (DMA). Conveniently, a shape memory cycle (SMC) as mentioned in **Figure 1** is performed in order to determine R_f and R_r, the changes of sample dimension are manually measured, and R_f and R_r are calculated with the measured data using Eqs. (1) and (2), respectively. However, a mechanical testing equipment with a climate chamber (possess heating and cooling facilities) is the best choice to evaluate ε_{max} by elongating the testing sample to its failure at T_{trans}. In addition to that, the SMC using such thermo-mechanical analyzer can precisely evaluate different shape memory parameters such as R_f, R_r, σ_{max}, and shape recovery rates. This method can accurately record the time progress of temperature, stress, and

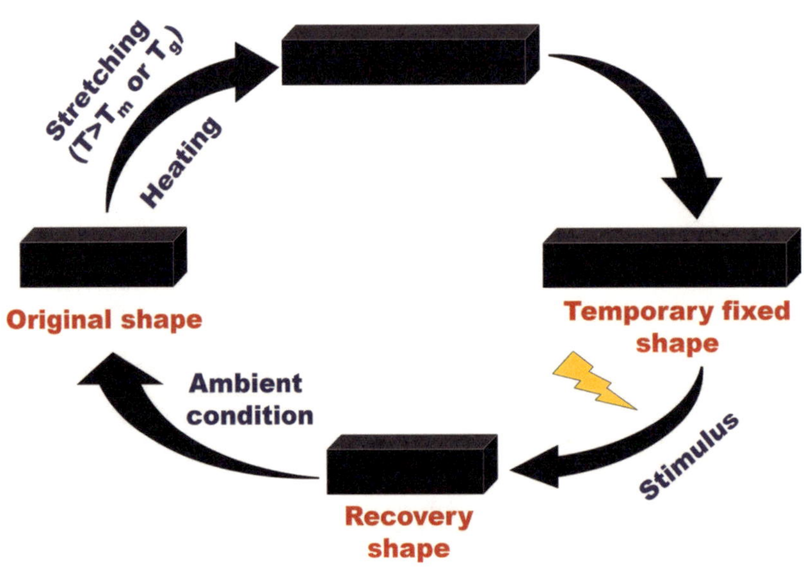

Figure 1. Thermomechanical cycle of SMPs.

strain. An example of a typical SMC is shown in **Figure 2a**. Instead, the SMC may also be demonstrated in a three-dimensional (3D) diagram as shown in **Figure 2b**. In this 3D diagram, the three axes are temperature, strain, and stress. Especially, time information is absent there. This absence does not impede the determination of R_f and R_r. Basically, the use of such a 3D diagram is very well known in the literature. The absence of time information may be moderately

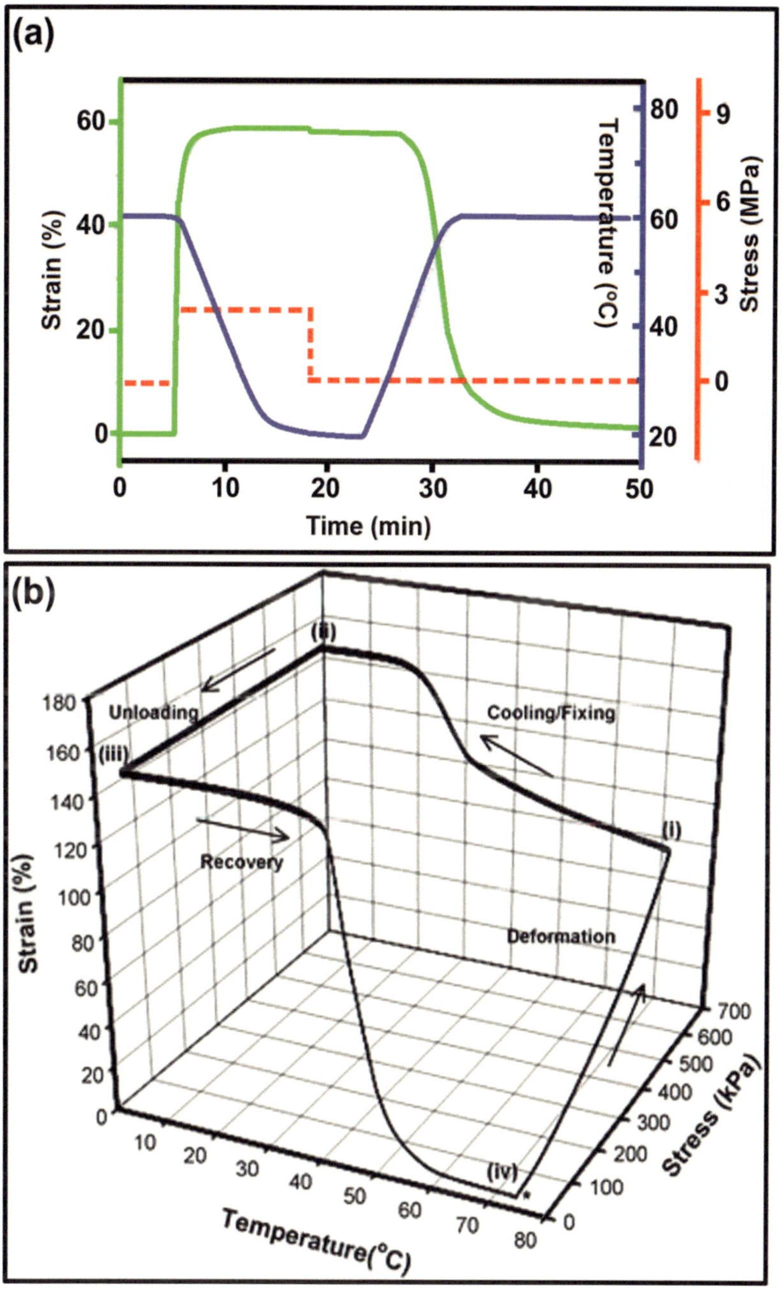

Figure 2. Shape memory cycle test: (a) 2D diagram (reproduced with permission from Ref. [2]) and (b) 3D diagram (reproduced with permission from Ref. [3].

unfavorable for a more sophisticated experiment of SMC. For example, if the testing sample is annealed under a constant stress at any stage of the experiment, the information about the annealing time and the strain reaches equilibrium or not during the annealing process would not be known. The rapid strain recovery rate V_r can also be calculated from the strain curve in the recovery portion of the SMC (**Figure 2a**), the time derivative of the strain as defined in Eq. (3)

$$V_r = \frac{\delta\varepsilon}{\delta t} \times 100\% \tag{3}$$

4. Mechanism of shape memory behavior of SMPs

It is well known that the segmental motions of the polymer chains ceased on cooling the polymer below T_g, but the motions start dramatically above this T_g. Basically, the polymers are transformed from a glassy state to a rubbery-elastic state during this transition. In this state, if uniaxial stress is applied for a short period of time, then the entanglement of polymer chains prevents a large-scale movement of chains, resulting in the storage of entropic energy. However, if the application of stress occurs over a longer period of time, the relaxation process can take place and that causes chain slippage and bulk flow of polymer chains. Thus, the reversible macromolecular deformation can be achieved by using network chains as a kind of molecular switch. At a certain dose of external applied energy (stimulus), the chains are flexible at a temperature above the T_{trans}, whereas their flexibility below T_{trans} is limited. Thus, the freezing of the molecular motion of the amorphous zone or the crystallization of the crystalline zone of the polymers prevents the molecular chains from immediately reforming the coil-like structures and instinctive recovery of the original shape, that is, programmed shape is fixed. Therefore, the stability of molecular orientation depends on the strength of interaction between the macromolecular segments and on the conformations of the chains constituting a polymer. The stretching of molecular chains leads to a drop in entropy, which can be compensated for by the cooling process, where the internal energy is decreased. Again, when the system is heated, the oriented polymer chains are softened from their glassy state or melted from the crystals, and thereby molecular chains relax the orientation to form more stable, coiled conformations. Such relaxation or shrinkage of the molecular chains caused shape recovery [23]. Therefore, the elastic strain energy produced during the deformation process is the driving force for shape recovery in SMPU. The molecular mechanism of shape memory behavior of SMPUs is shown in **Figure 3**.

At a high temperature, the rubber modulus of SMPU is lower, which makes the orientation of SMPU chains more feasible. So the deformation is much easier at high temperature. On the contrary, deformation at a lower temperature is tougher as glassy state modulus of the SMPU is high. However, the orientation of SMPU chain will withstand at a higher degree due to the slowdown of the relaxation process. Therefore, higher glassy state modulus (E_g) will offer higher shape fixity during concurrent cooling and unloading, where a higher rubber modulus (E_r) will offer higher elastic recovery at a higher temperature. Shape memory effect can thus be described by mathematical modeling as follows [24]:

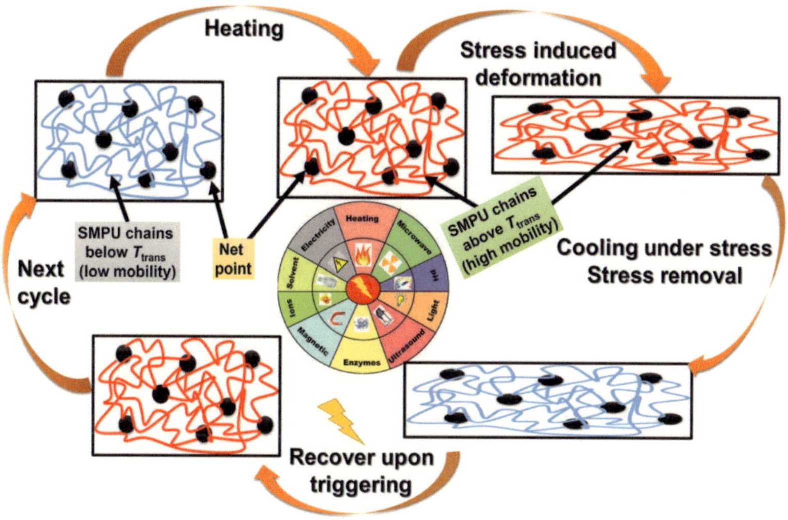

Figure 3. The molecular mechanism of the shape memory effect under different stimuli. Black dots: net points; blue lines: SMPU chains below T_{trans} (low mobility); red lines: SMPU chains above T_{trans} (high mobility).

$$R_f = (1 - \frac{E_r}{E_g}) \tag{4}$$

$$R_r = \frac{(1 - f_R)}{\left[\left(\frac{1 - E_r}{E_g}\right) f_\alpha\right]} \tag{5}$$

where f_R is the viscous flow strain and f_α is the strain when $T >>> T_s$. A high elastic ratio (E_g/E_r) offers easy shaping of SMPs at $T > T_s$ and provides a great resistance to deformation at $T < T_s$. The polymer should have thus greater E_g/E_r. The fixation of the temporary shape is caused by strain-induced crystallization and strain-oriented reorganization. The processing conditions of SMPs have also an effect on the shape memory behavior, as there may be a variation in modulus under different processing conditions. A significant variation of rubbery modulus in SMPU was observed when cooled at different rates. Further, the recovery ratio increased with the increase of deformation speed with decreasing maximum strain. T_s of SMPU could be tuned over a wide range of temperature from −30 to +100°C by using different structures and compositions of the components like diisocyanate, polyol (macroglycol), and chain extender. Further, the shape memory effect can be monitored by the proper choice of nature and the amount of reinforcing nanomaterials.

5. Criteria and designing protocol for a desired SMPU

As mentioned earlier, the presence of two different "segments" (a permanent or a hard segment and a switching or a soft segment, which can be influenced by the external trigger) is essential to obtain shape memory effect in SMPs. The hard phase is responsible for retaining the original shape of the whole SMP. This hard segment can be attained through chemical crosslinking in the polymer network, incorporation of interpenetrating networks or crystalline phases (**Figure 4**). The switching segment temporarily fixes the programmed shape by a glass transition, crystallization, a transition between different liquid crystalline phases, reversible covalent or non-covalent bonds (e.g., photodimerization of coumarin, Diels-Alder reactions, and supramolecular interactions).

The polymer must deform to a desirable extent on heating at a certain temperature without any other change of its performance. This certain temperature is known as switching temperature (T_{sw}), which may be a T_g or a T_m of switching segments of the polymers. Here, it is necessary to mention that if the T_{sw} is nearer to the human body temperature, then the polymer is very useful for biomedical applications and T_{sw} as T_m is preferable over the T_g, as the former is a sharper physical property than the latter. Upon reheating above the T_{sw}, the oriented chains in case of an amorphous polymer or a crystalline polymer in the network restore the random coil conformation resulting in a macroscopic recovery of the original shape. So they have two different segments or phases in the structures: one is a net point or hard or fixed phase and the other is a soft or reversible or switching segment. Thus, the hard segment serves as a pivoting point for shape recovery and the soft segment could mainly absorb external stress applied to the polymers. They should also possess highly strain fixity rate and strain recovery rate.

The programming like applied strain or deformation rate and extent are also influenced by the shape memory effect of the polymers. The recovery ratios increased with an increasing deformation speed and with a decreasing maximum strain. Thus for good shape recovery,

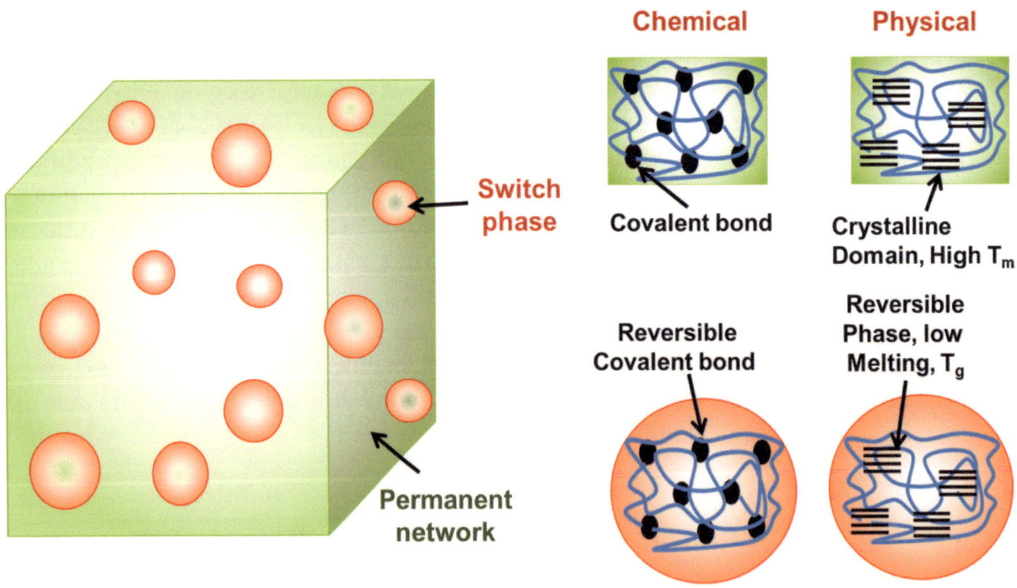

Figure 4. General structure of SMPU containing a "permanent network," which controls the shape and a "switch phase," which can be triggered by external stimuli.

deformation speed must be fast, while overall deformation should be low. The recovery stress of SMPs increased with the decrease of stretching rate, an increase of stretching temperature, and stretch ratio. The recovery stress of SMPU largely depended on the degree of interfacial interaction. The effective SMPs can be obtained when the hard segments retain the whole shape through inter-/intra-attractions (H-bonding or dipole-dipole interaction) in SMPs chain and the soft segments freely engross external stress by extending and unfolding the molecular chains. If the interactions in the hard segments are broken at high stress, the shape memory effect will be lost and the original shape cannot be restored. Hence, the controlling of composition and the structure of hard segments and soft segments are very much required to obtain desired shape memory effect.

SMPs are designed in such a way so that a large change in elastic modulus above and below T_g of the switching or amorphous phase will occur. They also have micro-Brownian movement of the chain molecules at a temperature above T_g, though rubber elasticity will present within T_g to T_m range due to limited molecular motion of the frozen or crystalline phase. Below T_g, the deformation is fixed due to the frozen Brownian motion of the chains. The shape can be recovered by reheating at an elevated temperature due to the recovery of the elastic force or the strain generated during the deformation.

For crystalline SMPU, T_m can be taken as at T_s instead of T_g, as a large change in elastic modulus above and below T_m of the soft segment is possible. Further, thermoplastic SMPU with the mesogenic unit in the structure has a high value of modulus compared with conventional SMPU [25]. Crystalline state, glassy state, entanglement network, or crosslinking network can be used as a fixed structure memorizing the original shape, that is, frozen phase, the reversible phase must have a large drop-in-elastic modulus on heating at T_s. The selection of T_s depends on the temperature at which the elastic modulus is the highest.

From the study of shape memory effect with the molecular structure, it has been observed that high crystallinity of the soft segment region at room temperature is a necessary prerequisite for SMPU [26]. Thus, the most important thing to be required for developing SMPU is to achieve maximum crystallization and stable hard segment domains. To achieve this, various attempts have been made. These include incorporation of the mesogenic unit such as 4,4′-bis (2-hydroxy ethoxy) biphenyl [27], ionic group [28], long alkyl chain [29], and so on. Further, the studies of Kim et al. suggested that large ratio of E_g to E_r is essential for obtaining good shape memory effect [29]. This can be easily achieved by using crystalline soft segments, for example, polycaprolactone diols; mesogenic moiety, for example, hydroxy-substituted biphenyls; and ionomers, for example, dimethylol propionic acid as the components of SMPU.

6. Application of shape memory polyurethane

SMP has myriad of advanced applications in different fields from daily life uses like textile to biomedicals. These applications include artificial organs like muscles, smart catheter, drug-delivery systems, biosensors, auto-repairing automobile parts, smart textile, self-healing materials, smart suture, and so on. In this section, we described the detailed application of SMPU in the different advanced areas.

6.1. Industrial applications of SMPU

SMPUs are already widely used as heat-shrinkable polymer tubings, films, and so on. The utilization of SMPU provides easier processing compared to other used polymers in such application [30]. As a consequence, these materials found a wide range of applications, for instance as a safety tag [31] or as a self-deploying chair [32].

6.2. Biomedical applications of SMPU

SMPs are also well suitable for the use in different biomedical applications, even though several requirements must be addressed and a range of problems must be overwhelmed for advanced application in this field [33]. For example, one major issue with thermo-responsive SMP is the heating of it inside the human body. Various approaches are developed to overcome this problem. One route is the use of noncontact triggering stimulus such as infrared (IR), lasers, and so on [34, 35]. These can heat SMP inside the body at the accurate location. Another way is the incorporation of magnetic nanoparticles in SMP. This magnetic nanoparticle can be triggered by an external magnetic field for the selective heating of the SMP. Biocompatible and nontoxic nature of SMP is also a crucial concern for biomedical applications. In this regard, it is pertinent to mention that the several SMPUs are developed, which are biocompatible. Generally, PCL, PEG, and polylactic acid-based SMPU are found to be nontoxic and biocompatible [36–38]. Besides these academic studies, DiAPLEX is a commercially available SMPU which also showed biocompatibility. Such biocompatible SMPU can be utilized in several biomedical applications, such as endovascular devices (clot-removal devices, aneurysm occlusion devices, and vascular stents), repair of cardiac valves, tissue engineering, orthopedics, orthodontics, endoscopic surgery, kidney dialysis, photodynamic therapy, and so on.

Thermo-responsive SMP-based mechanical clot extraction devices to treat ischemic stroke was reported by Maitland et al. [39]. Their fabricated catheter is photothermally activated, so it can easily remove the clot and finally relieve of the ischemia. Moaddeb and coworkers invented SMPU-based devices for treating heart failure patients suffering from various levels of heart dilation [40]. Such heart dilation is treated by reshaping the heart anatomy with the use of SMPU. The concept of biodegradable thermo-responsive SMP sutures was showed by Lendlein and Langer [41]. The suture was fabricated using oligo(ε-caprolactone)diol-based SMP. An abdominal wound in a rat was loosely sutured using the SMP fiber, and then heated to body temperature to achieve wound closure (**Figure 5a**). SMPU provides an alternative to traditional materials used for the treatment of dental malocclusions. Also, SMP arch wire in orthodontic braces for aligning teeth is more aesthetically appealing than a traditional metallic arch wire. These features were studied by Jung and Cho [42]. They used extruded SMP wire, which was attached to stainless steel brackets bonded to teeth in a dental model. When heated, the teeth slowly moved into alignment (**Figure 5b**).

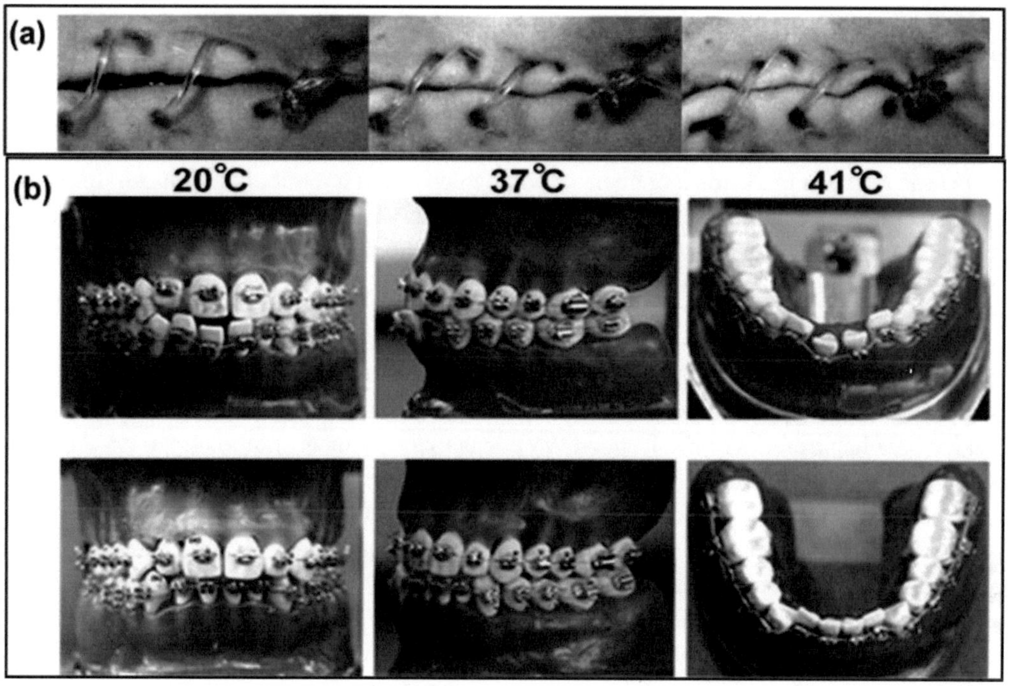

Figure 5. (a) Biodegradable SMP suture for wound closure. The photo series from the animal experiment shows the shrinkage of the suture as temperature increases (reproduced with permission from Ref. [41]) and (b) photographs of the orthodontic appliance (top) before and (bottom) after treatment. The movement of the misaligned teeth due to a lateral force originating from the shape recovery of the SMP arch wire is seen (reproduced with permission from Ref. [42]).

Tissue engineering is one of the large application areas of SMPs. The introduction of biodegradable SMPU urged the study of their usage for minimally invasive tissue engineering. Usually, tissues can be grown on SMPU-based scaffolds and incorporated inside the body through minimally invasive techniques (e.g., catheter). The scaffold is implanted to initiate the repair or reconstruction of tissues or organs in the affected area of the body. The SMPU-based-implantable embolic devices and stents demonstrated potential endovascular tissue engineering applications. Such biodegradable SMPU scaffolds can also be applied in pharyngeal mucosa reconstruction, bone regeneration, and organ repair. Different research groups have investigated the use of thermally responsive SMPUs as an extracellular matrix for in situ growing of various tissues. In this context, Rickert et al. reported the growth of cells on a biodegradable PCL-based SMP [43]. Rat pharyngeal cells are grown on a porous and smooth surface of the SMP to study the prospect of reconstructing the mucosa of the upper aerodigestive tract. Neuss et al. also investigated the cell-growing behavior of human mesenchymal stem cells, human mesothelial cells, rat mesothelial cells, and L929 mouse fibroblasts, on a similar PCL-based SMPU [44]. They found that mesothelial cells create an anti-adhesive surface layer, which may support abdominal repair or regeneration. Mesenchymal stem cells, the precursor cells of bone, fat, cartilage, and muscle, may support bone regeneration and the construction of adipose tissue. Furthermore, they also found that the use of heat as a stimulus did not affect the majority of adherent cells. SMPU/carbon nanotube composites also showed good MG63 cell differentiation ability, which showed its potentiality as an alternative biomaterial for bone regeneration in a comprehensive manner [45].

6.3. Textile application of SMPU

SMPU can be made in the form of fibers (macro-, micro-, and nanofibers), solutions, films, and foams for textile and apparel applications, such as non-woven materials, coatings, finishing, lamination, weaving, and knitting [46–49]. Various methods such as wet spun, melt spun, dry spun, and electrospun are used for the production of SMPU fiber [50–53].

Shape memory films and foams have a number of applications in laminated smart fabrics [54, 55]. The functions of SMP films applied to textiles include waterproofing, water vapor permeability (WVP), seam sewing, crease recoverability, and crease fixing [56, 57]. SMPU has a potential for making breathable fabrics due to its good WVP sensitivity. Jeong et al. studied the WVP properties of SMPUs, and breathable fabrics were invented by coating the SMPU membranes on a fabric substrate [26]. Mondal and Hu also designed SMPU-coated fabrics, which abruptly increased WVP properties at room temperature (T_r) compared to low temperature [58]. These results suggested that the breathable textiles require possessing a high WVP at higher temperatures and a low WVP at lower temperatures. Additionally, Mondal and Hu incorporated a small percentage of carbon nanotube in SMPUs and then coated on cotton. They reported that the fabricated cotton fabric provided excellent UV protection, along with a required WVP and wearing coziness [58]. In addition to that, Chen et al. also investigated adjusting the size and shape of the free-volume holes in a fabricated membrane to control the WVP by adjusting the temperature [59]. Further, Mondal and Hu attempted to find the influence of hydrophilic groups and crystalline soft segments on the WVP of SMPU films [60]. They found that the WVP increased with the increase of PEG due to the enhancement of hydrophilicity. However, PCL or polytetramethylene glycol-based SMPUs have low WVP because of the increased interaction among the polymer chains.

Hu et al. also fabricated a fabric-based thermoelectric generator using coating of waterborne PU composite on yarn [61]. They reported that this coated fabric showed satisfactory thermoelectric performance and good processability. In addition to that, medical stockings are also fabricated using SMPU for the treatment of chronic venous disorders [62–64]. This SMPU-based stocking allows controlling or managing the pressure exerted in a wrapped position and also produces extra pressure (up to 50%) by simply heating the stocking. This type of stocking possesses a great potential to overwhelm the restriction of conventional stockings. It can be used as a smart wound-care product, during the course of compression therapy. Self-healing textiles are also developed using SMPU. In this context, Hu et al. fabricated stimuli-responsive fiber using SMPU which showed 94% healing efficiency [65].

6.4. Anti-counterfeiting application of SMPU

SMP security label film is made from SMPU. SMP security label film can store embossed logo/text shape information in the synthetic-paper-like film, and release these information when exposed to stimuli (**Figure 6**). This kind of SMP security label is just not only an exclusive anti-counterfeit label but also a tamper evidence label [66]. These labels are used for security label, tamper evidence, security packaging, security ticket, and so on.

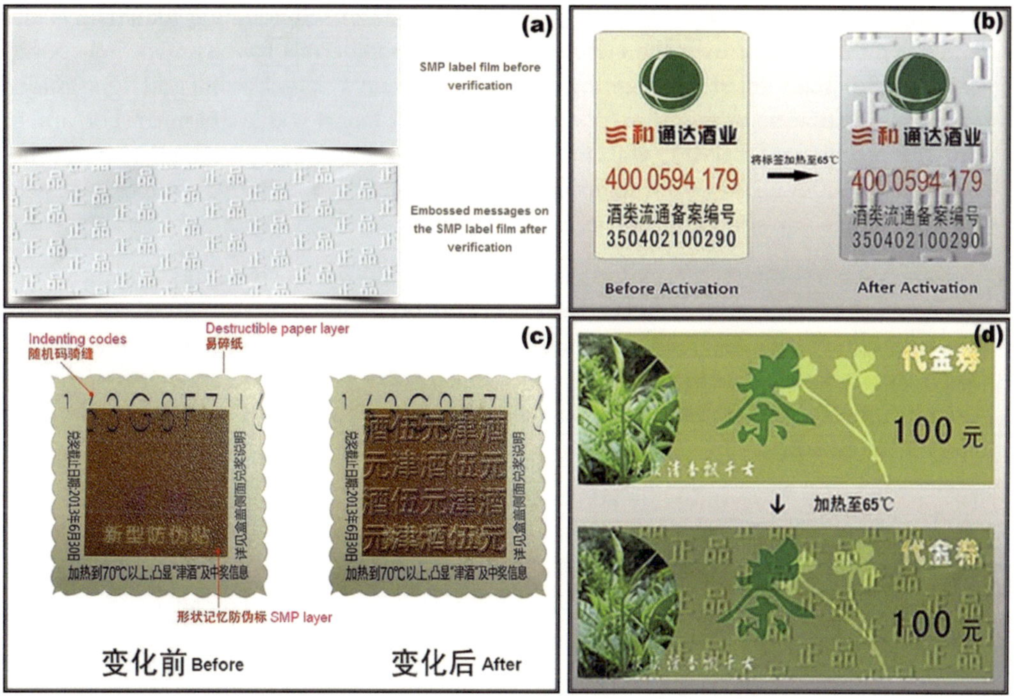

Figure 6. SMP in the anti-counterfeiting application (a) and (b) security label film, (c) tamper evidence label, and (d) security ticket. Adapted from Ref. [66].

7. New development and future trend

Even though an enormous progress in the field of SMPU is already done, several future directions and related challenges still remain, which may be considered in future research:

a. The fabrication of medical devices based on biocompatible and biodegradable SMPU is of great interest owing to the important impact on human health. Various materials are developed such as self-tightening sutures, scaffolds for tissue engineering, drug-delivery systems, implants for minimally invasive surgery procedures, synthesis of protein-polymer or DNA-polymer conjugates for therapeutic self-retractable and removable stents, and so on. Here, the main challenges of concern are the rate of degradation of SMPU (enzymatic or hydrolytic), their degree of toxicity, mechanical solicitation, and so on.

b. Three-dimensional printing of smart materials is another attractive field to be developed imminently. The change in the shape upon triggering of external stimuli provided a concept for the development of unique "four-dimensional (4D) printing" procedure. This may pave a way for the production of actuators to develop self-evolving structures, soft robotics, anti-counterfeiting system, and 4D bioprinting materials. Some technological and design restrictions are still unsolved such as inadequate choice of properly usable polymer, the existence of microstructural defects in SMPU, and materials real-time adapting; such problem can be addressed in future.

c. The design of SMPs-based self-cleaning, self-healing, and self-adapting materials is also gaining much interest over the last few years. These materials have shown their potential for the elaboration of load-bearing aircraft components, self-cleaning and light-guided windows, flexible solar modules (polymer solar cells), smart textiles, bionic robot, and so on. The production of SMPU with improved durability and good mechanical properties will further improve the performance in the field. In this regard, SMPU composite may be a choice to develop such smart materials.

d. The energy-harvesting SMPs (solar energy) and chemical-responsive SMPs (chemical reaction energy) also carved a distinct attention of both academic and commercial researchers. The molecular-, gas-, or sound-detectable materials can be defined as forthcoming materials. In addition to that, wireless and remote-controllable SMPs are another class of devices to be produced using different magnetic and light-actuated composites and piezoelectric-containing materials.

e. Despite their multi-applicability and a wide range of properties, the direct transfer from the laboratory to industrial scale remains a challenge. In this context, the main problems are associated with the complexity of shape memory effect. These are affected by many factors such as the programming step and the triggering process parameters. In addition, quick and versatile manufacturing processes are also desirable, while focusing on the use of cost-effective and eco-friendly raw materials, and ensured profit gain, and so on.

Author details

Suman Thakur and Jinlian Hu*

*Address all correspondence to: jin-lian.hu@polyu.edu.hk

Institute of Textiles and Clothing, The Hong Kong Polytechnic University, Hung Hom, Hong Kong

References

[1] Meng H, Hu J. A brief review of stimulus-active polymers responsive to thermal, light, magnetic, electric, and water/solvent stimuli. Journal of Intelligent Material Systems and Structures. 2010;**21**(9):859-885. DOI: 10.1177/1045389X10369718

[2] Xie T. Tunable polymer multi-shape memory effect. Nature. 2010;**464**:267-270. DOI: 10.1038/nature08863

[3] Kunzelman J, Chung T, Mather PT, Weder C. Shape memory polymers with built-in threshold temperature sensors. Journal of Materials Chemistry. 2008;**18**:1082-1086. DOI: 10.1039/B718445J

[4] Meng Q, Hu J. A review of shape memory polymer composites and polymer composites and blends. Composites Part A. 2009;**40**(11):1661-1672. DOI: 10.1016/j.compositesa.2009.08.011

[5] Ji FL, Hu J, Li TC, Wong YW. Morphology and shape memory effect of segmented polyurethanes. Part I: With crystalline reversible phase. Polymer. 2007;**48**(17):5133-5145. DOI: 10.1016/j.polymer.2007.06.032

[6] Meng H, Li G. A review of stimuli-responsive shape memory polymer composites. Polymer. 2013;**54**:2199-2221. DOI: 10.1016/j.polymer.2013.02.023

[7] Sakurai K, Takahashi T. Strain-induced crystallization in polynorbornene. Journal of Applied Polymer Science. 1989;**38**:1191-1194. DOI: 10.1002/app.1989.070380616

[8] Thakur S, Karak N. Bio-based tough hyperbranched polyurethane-graphene oxide nanocomposites as advanced shape memory materials. RSC Advances. 2013;**3**:9476-9482. DOI: 10.1039/C3RA40801A

[9] Thakur S, Karak N. Multi-stimuli responsive smart elastomeric hyperbranched polyurethane/reduced graphene oxide nanocomposites. Journal of Materials Chemistry A. 2014;**2**:14867-14875. DOI: 10.1039/C4TA02497D

[10] Thakur S, Karak N. A tough, smart elastomeric bio-based hyperbranched polyurethane nanocomposites. New Journal of Chemistry. 2015;**39**:2146-2154. DOI: 10.1039/C4NJ01989J

[11] Sahoo NG, Jung YC, Yoo HJ, Cho JW. Influence of carbon nanotubes and polypyrrole on the thermal, mechanical and electroactive shape-memory properties of polyurethane nanocomposites. Composites Science and Technology. 2007;**67**:1920-1929. DOI: 10.1016/j.compscitech.2006.10.013

[12] Thakur S, Karak N. Castor oil-based hyperbranched polyurethanes as advanced surface coating materials. Progress in Organic Coatings. 2013;**76**:157-164. DOI: 10.1016/j.porgcoat.2012.09.001

[13] Hu J, Lu J, Zhu Y. New developments in elastic fibers. Polymer Reviews. 2008;**48**:275-301. DOI: 10.1080/15583720802020186

[14] Hu J, Yang Z, Yeung L, Ji F, Liu Y. Crosslinked polyurethanes with shape memory properties. Polymer International. 2005;**54**:854-859. DOI: 10.1002/pi.1785

[15] Salamore JC. Concise Polymeric Materials Encyclopedia. Boca Raton: CRC Press; 1999

[16] Madbouly SA, Otaigbe JU. Recent advances in synthesis, characterization and rheological properties of polyurethanes and POSS/polyurethane nanocomposites dispersions and films. Progress in Polymer Science. 2009;**34**:1283-1332. DOI: 10.1016/j.progpolymsci.2009.08.002

[17] Chattopadhyay DK, Raju KVSN. Structural engineering of polyurethane coatings for high performance applications. Progress in Polymer Science. 2007;**32**:352-418. DOI: 10.1016/j.progpolymsci.2006.05.003

[18] Lamba NMK, Woodhouse KA, Cooper SL. Polyurethane in Biomedical Application. Boca Raton: CRC Press; 1997

[19] Zhang C, Feng S. Effect of glycols on the properties of polyester polyols and of room-temperature-curable casting polyurethanes. Polymer International. 2004;**53**:1936-1940. DOI: 10.1002/pi.1571

[20] Bae JY, Chung DJ, An JH, Shin DH. Effect of the structure of chain extenders on the dynamic mechanical behaviour of polyurethane. Journal of Material Science. 1999;**34**:2523-2527. DOI: 10.1023/A:1004680127857

[21] Hepburn C. Polyurethane Elastomers. 2nd ed. London: Elsevier Applied Science; 1992

[22] Zhao Q, Qi HJ, Xie T. Recent progress in shape memory polymer: New behavior, enabling materials, and mechanistic understanding. Progress in Polymer Science. 2015;**49-50**:79-120. DOI: 10.1016/j.progpolymsci.2015.04.001

[23] Jeong HM, Ahn BK, Kim BK. Miscibility and shape memory effect of thermoplastic polyurethane blends with phenoxy resin. European Polymer Journal. 2001;**37**:2245-2252. DOI: 10.1016/S0014-3057(01)00123-9

[24] Yao Y, Zhou1 T, Qin C, Liu Y, Leng J. Styrene-based shape memory foam: Fabrication and mathematical modeling. Smart Materials and Structures. 2016;**25**:105031 (10pp). DOI: 10.1088/0964-1726/25/10/105031

[25] Xie T. Recent advances in polymer shape memory. Polymer. 2011;**52**:4985-5000. DOI: 10.1016/j.polymer.2011.08.003

[26] Jeong HM, Ahn BK, Kim BK. Temperature sensitive water vapour permeability and shape memory effect of polyurethane with crystalline reversible phase and hydrophilic segments. Polymer International. 2000;**49**:1714-1721. DOI: 10.1002/1097-0126(200012)49:12<1714::AID-PI602>3.0.CO;2-K

[27] Kim BK, Shin YJ, Cho SM, Jeong HM. Shape-memory behavior of segmented polyurethanes with an amorphous reversible phase: The effect of block length and content. Journal of Polymer Science Part B: Polymer Physics. 2000;**38**:2652-2657. DOI: 10.1002/1099-0488(20001015)38:20<2652::AID-POLB50>3.0.CO;2-3

[28] Zhu Y, Hu J, Yeung KW, Fan HJ, Liu YG. Shape memory effect of PU ionomers with ionic groups on hard-segments. Chinese Journal of Polymer Science. 2006;**24**:173-186. DOI: 10.1142/S0256767906001187

[29] Kim BK, Lee SY, Lee JS, Baek SH, Choi YJ, Lee JO, Xu M. Polyurethane ionomers having shape memory effects. Polymer. 1998;**39**:2803. DOI: 10.1016/S0032-3861(97)00616-2

[30] Huang WM, Yang B, Qing Y. Polyurethane Shape Memory Polymers. Boca Raton: CRC Press; 2012

[31] Anonymous TPU Newsletter. March 2012. Available from: www.bayermaterialscience.com

[32] de Smet C. Marcus Fairs Memories of the Future. 2012. Available from: www.dezeen.com/2012/10/25/noumenon-by-carl-de-smet/

[33] Small IVW, Singhal P, Wilson TS, Maitland DJ. Biomedical applications of thermally activated shape memory polymers. Journal of Material Chemistry. 2010;**20**:3356-3366. DOI: 10.1039/B923717H

[34] Small IVW, Wilson TS, Benett WJ, Loge JM, Maitland DJ. Laser activated shape memory polymer intravascular thrombectomy device. Optics Express. 2005;**13**:8204-8213. DOI: 10.1364/OPEX.13.008204

[35] Small IVW, Buckley PR, Wilson TS, Benett WJ, Hartman J, Saloner D, Maitland DJ. Shape memory polymer stent with expandable foam: A new concept for endovascular embolization of fusiform aneurysms. IEEE Xplore: IEEE Transactions on Biomedical Engineering. 2007;**54**:1157-1160. DOI: 10.1109/TBME.2006.889771

[36] Ping P, Wang W, Zhang P, Chen X, Jing X. Shape-memory and biocompatibility properties of segmented polyurethanes based on poly(L-lactide). Frontiers of Chemistry in China. 2007;**4**:331-336. DOI: 10.1007/s11458-007-0063-z

[37] Neuss S, Blomenkamp I, Stainforth R, Boltersdorf D, Jansen M, Butz N, Perez-Bouza A, Knüchel R. The use of a shape memory poly(ε-caprolactone)dimethacrylate network as a tissue engineering scaffold. Biomaterials. 2009;**30**:1697-1705. DOI: 10.1016/j.biomaterials.2008.12.027

[38] Yakacki CM, Lyons MB, Rech B, Gall K, Shandas R. Cytotoxicity and thermomechanical behavior of biomedical shape-memory polymer networks post-sterilization. Biomedical Materials. 2008;**3**:015010/1-9. DOI: 10.1088/1748-6041/3/1/015010

[39] Maitland DJ, Metzger MF, Schumann D, Lee A, Wilson TS. Photothermal properties of shape memory polymer micro actuators for treating stroke. Lasers in Surgery and Medicine. 2002;**30**:1-11. DOI: 10.1002/lsm.10007

[40] Moaddeb S, Shaolian SM, Shaoulian E, Rhee R, Anderson SC. Shape memory devices and methods for reshaping heart anatomy. 2007, US7285087

[41] Lendlein A, Langer R. Biodegradable, elastic shape-memory polymers for potential biomedical applications. Science. 2002;**296**:1673-1676. DOI: 10.1126/science.1066102

[42] Jung YC, Cho JW. Application of shape memory polyurethane in orthodontic. Journal of Materials Science: Materials in Medicine. 2010;**21**:2881-2886. DOI: 10.1007/s10856-008-3538-7

[43] Rickert D, Lendlein A, Peters I, Moses M, Franke R-P. Biocompatibility testing of novel multifunctional polymeric biomaterials for tissue engineering applications in head and neck surgery: An overview. European Archives of Oto-Rhino-Laryngology and Head & Neck. 2006;**263**:215-222. DOI: 10.1007/s00405-005-0950-1

[44] Neuss S, Blomenkamp I, Stainforth R, Boltersdorf D, Jansen M, Butz N, Perez-Bouza A, Knuchel R. The use of a shape-memory poly(ε-caprolactone)dimethacrylate network as a tissue engineering scaffold. Biomaterials. 2009;**30**:1697-1705. DOI: 10.1016/j.biomaterials.2008.12.027

[45] Das B, Chattopadhyay P, Maji S, Upadhyay A, Das Purkayastha M, Mohanta C, Maity T K, Karak N. Bio-functionalized MWCNT/hyperbranched polyurethane bio-nano-composite for bone regeneration. Biomedical Materials. 2015;**10**:025011. DOI: 10.1088/1748-6041/10/2/025011

[46] Hu J, Meng H, Li G, Ibekwe S. A review of stimuli-responsive polymers for smart textile applications. Smart Materials and Structures. 2012;**21**:053001. DOI: 10.1088/0964-1726/21/5/053001

[47] Hu J. Shape Memory Polymers and Textiles. 1st ed. New York, NY, USA: Cambridge: Woodhead Publishing; 2007. p. 339

[48] Hu J, Lu J. Smart polymers for textile applications. In: Aguilar MR, Román JS, editors. Smart Polymers and Their Applications. 1st ed. New Delhi: Woodhead Publishing; 2014. pp. 437-475

[49] Hu J, Zhu Y, Huang H, Lu J. Recent advances in shape-memory polymers: Structure, mechanism, functionality, modeling and applications. Progress in Polymer Science. 2012;**37**:1720-1763. DOI: 10.1016/j.progpolymsci.2012.06.001

[50] Hu J, Chen S. A review of actively moving polymers in textile applications. Journal of Materials Chemistry. 2010;**20**(17):3346-3355. DOI: 10.1039/B922872A

[51] Zhao H, Hu J, Chen S, Yeung L. Preparation of polyurethane nanofibers by electrospinning. Journal of Applied Polymer Science. 2008;**109**:406-411. DOI: 10.1002/app.28067

[52] Zhu Y, Hu J, Yeung L-Y, Liu Y, Ji F, Yeung KW. Development of shape memory polyurethane fiber with complete shape recoverability. Smart Materials and Structures. 2006;**15**:1385. DOI: 10.1088/0964-1726/15/5/027

[53] Meng Q, Hu J, Zhu Y, Lu J, Liu Y. Morphology, phase separation, thermal and mechanical property differences of shape memory fibres prepared by different spinning methods. Smart Materials and Structures. 2007;**16**:1192. DOI: 10.1088/0964-1726/16/4/030

[54] Meng G, Hu J. A temperature-regulating fiber made of PEG-based smart copolymer. Solar Energy Materials and Solar Cells. 2010;**92**:1245-1252. DOI: 10.1016/j.solmat.2008.04.027

[55] Ji FL, Zhu Y, Hu J, Liu Y, Yeung LY, Ye GD. Smart polymer fibers with shape memory effect. Smart Materials and Structures. 2006;**15**:1547. DOI: 10.1088/0964-1726/15/6/006

[56] Hu J, Lu J. Shape memory fibers. In: Tao X, editor. Handbook of Smart Textiles. 1st ed. Singapore: Springer Science; 2015. pp. 183-207. DOI: 10.1007/978-981-4451-45-1_3

[57] Jin GX. Effect of structure parameters on memory properties of thermo-sensitive shape-memory woven fabrics. Journal of Textile Research. 2010;**31**:40-44

[58] Mondal S, Hu J. Water vapor permeability of cotton fabrics coated with shape memory polyurethane. Carbohydrate Polymers. 2007;**67**:282-287. DOI: 10.1016/j.carbpol.2006.05.030

[59] Chen Y, Liu Y, Fan HJ, Li H, Shi B, Zhou H, Peng BY. The polyurethane membranes with temperature sensitivity for water vapor permeation. Journal of Membrane Science. 2007;**287**:192-197. DOI: 10.1016/j.memsci.2006.10.028

[60] Mondal S, Hu J, Yong Z. Free volume and water vapor permeability of dense segmented polyurethane membrane. Journal of Membrane Science. 2006;**280**:427-432. DOI: 10.1016/j.memsci.2006.01.047

[61] Wu Q, Hu J. Waterborne polyurethane based thermoelectric composites and their application potential in wearable thermoelectric textiles. Composites Part B: Engineering. 2016;**107**(15):59-66. DOI: 10.1016/j.compositesb.2016.09.068

[62] Kumar B, Hu J, Pan N. Smart medical stocking using memory polymer for chronic venous disorders. Biomaterials. 2016;**75**:174-181. DOI: 10.1016/j.biomaterials.2015.10.032

[63] Kumar B, Hu J, Pan N. Memory bandage for functional compression management for venous ulcers. Fibers. 2016;**4**(1):10. DOI: 10.3390/fib4010010

[64] Narayana H, Hu J, Kumar B, Shang S, Han J, Liu P, Lin T, Jia FL, Zhu Y. Stress-memory polymeric filaments for advanced compression therapy. Journal of Materials Chemistry B. 2017;**5**:1905-1916. DOI: 10.1039/C6TB03354G

[65] Li G, Meng H, Hu J. Healable thermoset polymer composite embedded with stimuli-responsive fibres. Journal of the Royal Society Interface. 2012;**9**(77):3279-3287. DOI: 10.1098/rsif.2012.0409

[66] Guangzhou Manborui Material Technology Co., Ltd. Unique Shape Memory Polymer Label & Applications [Internet]. Available from: https://www.china-mbr.com/ [Accessed: 04/07/2017]

Chapter 4

Measurement and Numerical Modeling of Mechanical Properties of Polyurethane Foams

Michal Petrů and Ondřej Novák

Additional information is available at the end of the chapter

http://dx.doi.org/10.5772/intechopen.69700

Abstract

This chapter focuses on determination of mechanical properties of polyurethane foams applied in automotive industry and medicine. These materials have strong nonlinear viscoelastic behavior that is time dependent. A comprehensive description of their characteristics is very tough and difficult. Mechanical properties can to some extent be studied using mathematical models that need to be verified with measurements. This chapter describes selected mathematical relationships, rheological models, and also numerical simulations that can approximately describe mechanical properties. Mechanical properties of polyurethane foam are influenced by internal and external structure, shape and size of the cells, filling volume, and properties of the used polymer. Studied mechanical properties are contact pressures, stress distribution, and the dependence of stress on the strain rate.

Keywords: PU foam, stiffness, dynamical tests, contact pressure, FEM

1. Introduction

Mechanical properties of polyurethane foams compressed during fast and dynamic states are among others affected by a friction of the cellular structure and the air contained in the cells [1–3]. However, this does not significantly contribute to a compressive force in certain range of the deformation as reported in Ref. [4]. On behavior under compression, loading a significant influence has mechanical and physical properties of the PU foam such as a geometry, thickness, or density, which is explained by [5]. A decrease of the thickness of the comfort stuff from polyurethane foam brings desired weight reduction, but it reduces the total damping, and on the contrary, the total stiffness increases. Therefore, the trend in automotive industry has been a development of low density polyurethane foams with lower bulk density having better mechanical properties under compression than commonly used PU foams [2]. Another solution that can change mechanical properties is a vertical layering of polyurethane foams having different physical and mechanical properties. But it does not bring an expected improvement of characteristics corresponding with a composite behavior [6, 7], which may be due to the fact that only the layering of polyurethane foams does not bring the desired synergistic effect as stated by Ref. [8]. Also, a significant improvement in energy savings and vibroisolation characteristics or other parameters such as a permeability of PU foam is not achieved. The improvements are reflected only in reduced values of contact pressures from the load body [9]. The mechanical behavior of polyurethane foams can be considered substantially non-linear, with a large viscoelastic deformation, relaxation, and recovery of the structure.

2. Analysis of the properties of selected PU foam samples

Mechanical, chemical, and physical properties together with the experimentally identified structural properties of samples of different densities ρ_{PU} PU foams have been published in Refs. [1, 2, 5]. The structure of the polyurethane foam is formed by a chemical process polyaddition from alcohols with two or more hydroxyl groups and isocyanates. Isocyanate reacts with water to form carbon dioxide that creates cell structure of polyurethane foam. According to the applied type of ingredients and their ratio, PU foam can be divided into soft, moderately stiff, rigid, or hard. The flexibility of the cell structure is among others dependent on its density $\rho_{PU} = weight/volume$, which for comfort application is in the range from 10 to 100 kg m^{-3}. Samples of polyurethane foams can be characterized with low-permeable envelop, which arises as a consequence of heat removal from polyurethane with a mold wall. The internal cell structure is characterized by a distribution curve of the cell diameters, and it is significantly more porous, which, for porous structure, can be expressed by a dimensionless quantity Ψ in accordance with Eq. (1). The relationship describes the ratio of the polyurethane structure volume and total volume of structure, which is important for obtaining of the parameter named as a packing density. A number of porous cells and the connecting edges (edge connecting air cells) are significantly influenced by the diameter of cells as indicated in Ref. [1]. He also states that the structure of PU foam created by combining of individual cells is the macroscopic homogeneous system and regardless of the variability of cell diameters, and therefore it can replace a continuum or rheological models. In terms of a deformation mechanism, the behavior of foams can be characterized by the following aspects: during foam compression, the air escapes from the cell, the cell walls are bent, and from a certain phase, cell walls are in a contact with specific friction. During unloading, the air is sucked again into the structure. Therefore, for the fast compression of the cellular structure of the foam, the mechanical properties depend especially on the amount of air and the breathability of porous cells and thus on strain rate $\dot{\varepsilon}(t)$. Already in 1970 in Ref. [10], it was published that the rate of air cells has an influence on the value of an energy dissipation $\vartheta(t)$, which PU foam can absorb. From the viewpoint of the mechanical properties, polyurethane foams are nearly isotropic viscoelastic materials. It has been published in 1987 by Ref. [11]. They state that during tri-axial test, wherein the sample is simultaneously loaded in three main directions of the basic coordinate system (*X*, *Y*, and *Z*), approximately same course of the loading curves is obtained. They differ only in constant. As a result, deformation/strain ε in the main axis of the load and a volumetric deformation γ can be expressed with Eqs. (2) and (3). The results of analysis of the structure of PU foam samples with dimensions of 100 × 100 × 40 mm, which were numbered 1–6, are shown in **Table 1**. Air volume in the analyzed samples reached 96.5 ± 0.5%, wherein the parameter of the packing density Ψ was from 0.033 to 0.034. It is due to the density of pure polyurethane polymer (for the comfort stuff, it is from 1200 to 1500 kg m^{-3}). Depending on the increasing amount of polyurethane in the PU foam sample, the air volume decreases. This fact can be illustrated with a parametric graph (**Figure 1**).

$$\Psi = \frac{V_{polym}}{V_{air}}, \tag{1}$$

where Ψ [−] is the parameter of the packing density of PU foam, V_{polym} is the volume of PU foam [m^3], and V_{air} [m^3] is volume of the air.

Sample	Density [kg m^{-3}]	Areal weight [g m^{-2}]	Air content [%]	Ψ [−]	Average size of cells (inner structure) [μm]	Average size of cells (outer structure) [μm]
1	48 ± 0.01	1920 ± 0.40	96.0 ± 0.80	0.032	544 ± 37	687 ± 51
2	51 ± 0.26	2045 ± 5.20	96.5 ± 0.92	0.034	446 ± 43	562 ± 48
3	50 ± 0.16	2003 ± 3.20	96.6 ± 0.67	0.033	478 ± 26	601 ± 63
4	47 ± 0.33	1887 ± 3	96.8 ± 0.70	0.031	593 ± 52	690 ± 39
5	50 ± 0.12	2002 ± 0.40	96.6 ± 0.27	0.034	462 ± 27	613 ± 53
6	49 ± 0.39	1966 ± 7.13	96.7 ± 0.22	0.033	455 ± 45	625 ± 38

Table 1. Parameters of tested PU samples.

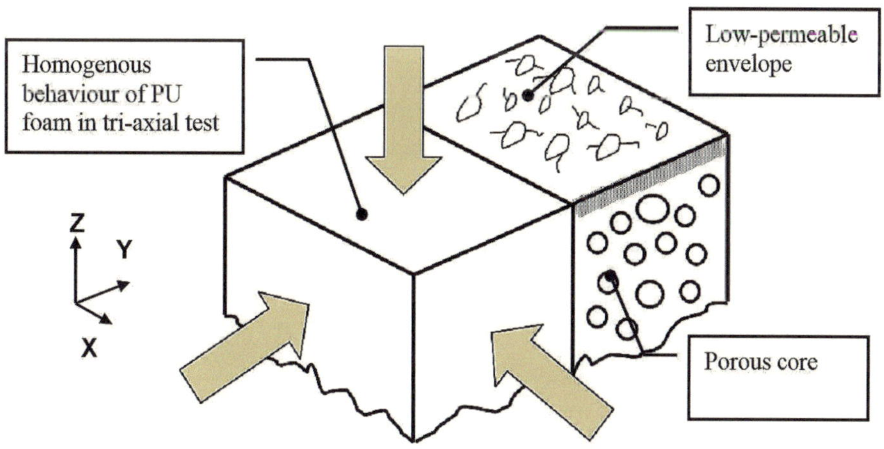

Figure 1. Characteristic structure of PU foam.

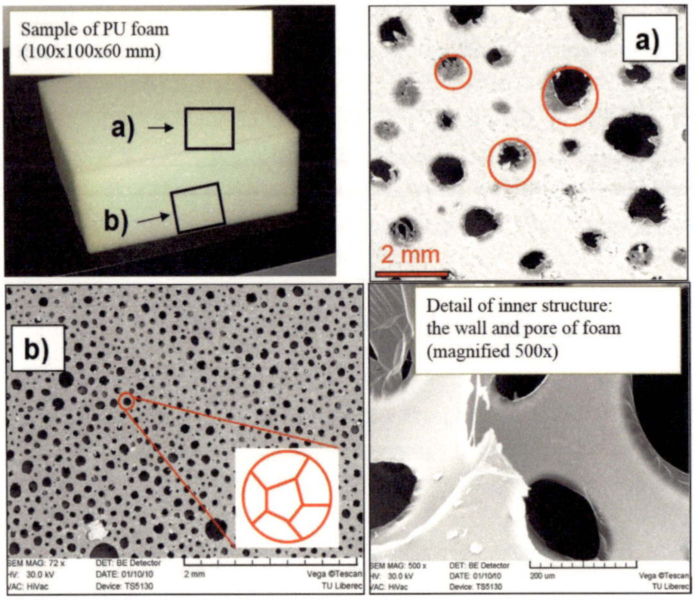

Figure 2. (a) Inner structure (low-porous envelop). (b) Inner structure of PU foam with characteristic shape of cell (left) and detail of inner structure (right).

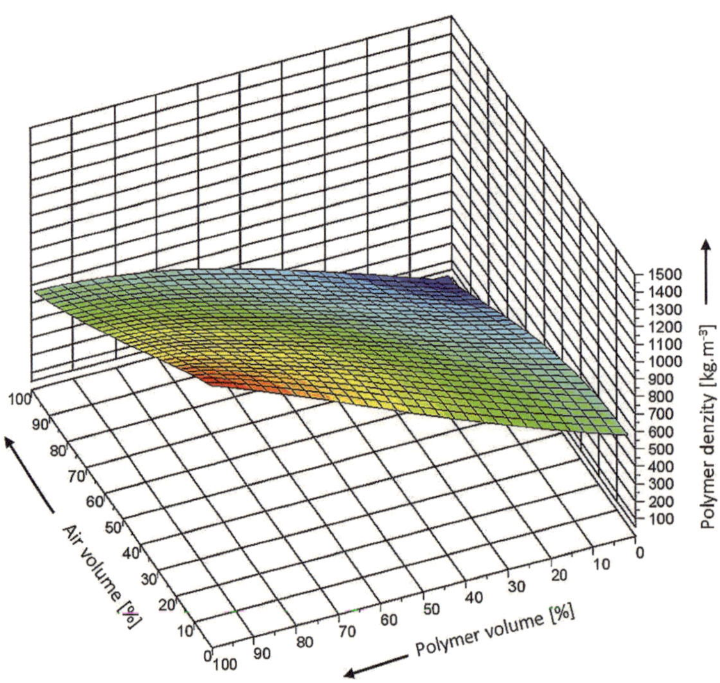

Figure 3. The volume of air and material in the structure of polyurethane foams depending on the specific weight of pure polymer.

$$\Gamma = \frac{\Delta V}{V_0} = \frac{V_0 - V_{comp}}{V_0} = \frac{\left(V_{air} + V_{polym}\right) - V_{comp}}{\left(V_{air} + V_{polym}\right)} = 1 - \Phi, \tag{2}$$

$$\varepsilon = \frac{L_0 - \delta}{L_0} = 1 - \left[1 - \frac{V_0 - V_{comp}}{V_0}\right]^{1/3} = 1 - [1 - \gamma]^{1/3}, \tag{3}$$

where $\Gamma[-]$ is total volume strain of PU foam, V_{comp} [m^3] is compressed volume, V_0 [m^3] is undeformed volume, V_{polym} [m^3] is volume of the polymer in PU foam, V_{air} [m^3] is volume of air in cells, Φ [−] is ratio of compressed and uncompressed volume, ε [−] is strain, δ [mm] is value (length) of compression, and L_0 [mm] is origin undeformed length (**Figures 2** and **3**).

3. Measurement of properties of selected samples during static compression

Mechanical properties of selected samples of PU foams show a stress dependent on the strain rate, which is accompanied by a change of stiffness. The stiffness of the sample can be experimentally determined as a slope of the tangent of the force depending on compression or deformation. The most significant change of K is in the initial phase (area no. 1, **Figure 13**). The material damping η_t is a mechanical variable that is difficult to measure. It can be approximated for example by energy dissipation in a hysteresis curve. But it should be understood that the obtained values will vary depending on the strain rate and the geometry of the loading body. Generally, two foams with the same density may not have the same stiffness. Comparative parameter may be contact pressures and the transmission characteristic that the value of the resonant frequency determines. There are numerous of methodologies describing the mechanical properties of PU foam, but they are not standardized. Therefore, own measurement methodology was designed and implemented for a measurement of properties under static and dynamic compression.

3.1. Determination of mechanical properties of samples of PU foams under static compression

For obtaining mechanical properties of selected materials under static or quasi-static compression, the measurements were made on samples compressed with rigid steel plate with dimensions of 200 × 200 × 50 mm. The specimen having areal dimension 100 × 100 mm was placed on the rigid support. Samples change mechanical properties with the thickness that is reflected in the changing of the stiffness and damping. These properties have measured on the PU foam with thickness 60, 40 and 20 mm (sample No. 3). For the tests (Figure 4) an universal testing machine Labortech 2.050 was used. A tension load cell with loading capacity 1 kN was placed on a moving part of the measuring device. For the measurement, up to 50% deformation in accordance with the standard DIN 54 305 was applied. This standard is used for quasi-static compression tests of bulky fibrous structures, PU foam, and similar materials. The strain rate was set 60 mm min^{-1}. After deformation, achieving the unloading phase up to 0% deformation

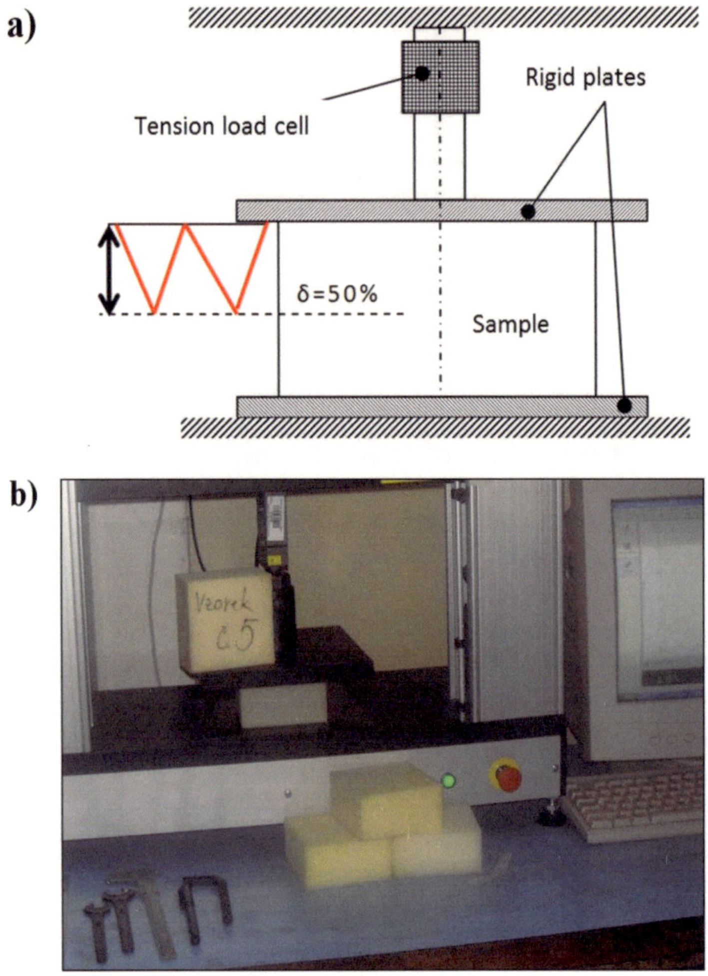

Figure 4. Determination of mechanical properties of samples of polyurethane foams during static compression: (a) scheme, (b) realization of the measurement.

follows. This represents one cycle that is four times repeated. A loading signal has a triangular wave shape. During the test, the force depending on compression is recorded. The results of PU foam samples with different thickness are shown in **Figure 5**. The results have confirmed that with decreasing thickness, the force required to compression of the sample to the desired deformation increases. It is also seen that between loading and unloading cycle is a hysteresis, and between first and second cycles, a significant loss of force occurs (relaxation of the material). The stiffness of samples was measured at the 5th cycle (**Figure 6**). The stiffness of samples with thickness 60 and 40 mm in the range between 30 and 50 % exhibits a difference about 20 % (2000 N/m), while highest force at 50% deformation differs only about 30 N. A sample with a thickness of 20 mm exhibited an increase in strength of 80 N in comparison with a sample thickness of

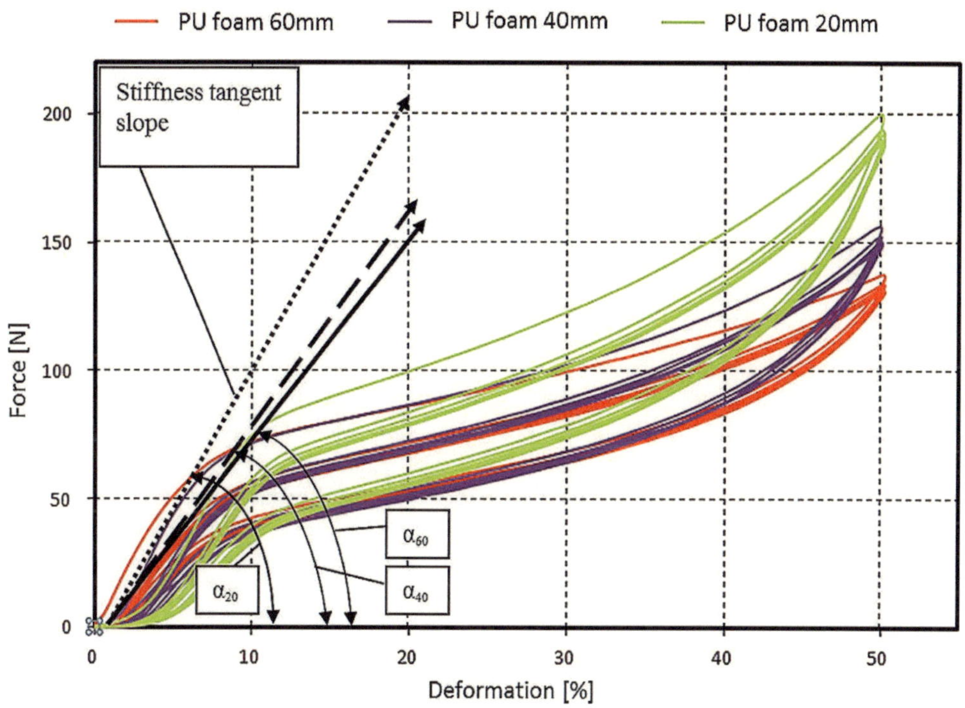

Figure 5. Dependence of force on the deformation of PU foam samples with dimensions 100 × 100 × 60, 40 and 20 mm during cyclic compression.

60 mm, and increase in strength of 50 N in comparison with a sample having a thickness of 40 mm. However, the stiffness of the sample having a thickness of 20 mm compared with the samples having a thickness of 40 and 60 mm increased about approximately 11,000 N/m. There can be applied the inequality $\alpha_{60} < \alpha_{40} < \alpha_{20}$, that describes tangent slope of initial stiffness of the cellular structure of a given thickness. It can be concluded that the targeted reduction in the thickness of the polyurethane foam (the idea of new seat and back car seats design) is not useful for the quality of the seating. It is not suitable especially for safety reasons. For example, currently manufactured headrests, where the structure of comfortable filler is made of PU foam having the thickness less than 20 mm exhibit at high strain rate significant hardening of the structure [3].

3.2. Determination of mechanical properties of PU samples during dynamic compression

Mechanical properties during dynamic compression relate to the ability of the material to dampen incoming vibrations with a given frequency and amplitude. It is caused by the reorganization of the structure, in this instance cellular, in which entering mechanical energy is being transformed to heat in a short time interval. Great amount of the dissipated mechanical energy $\vartheta(t)$ that was

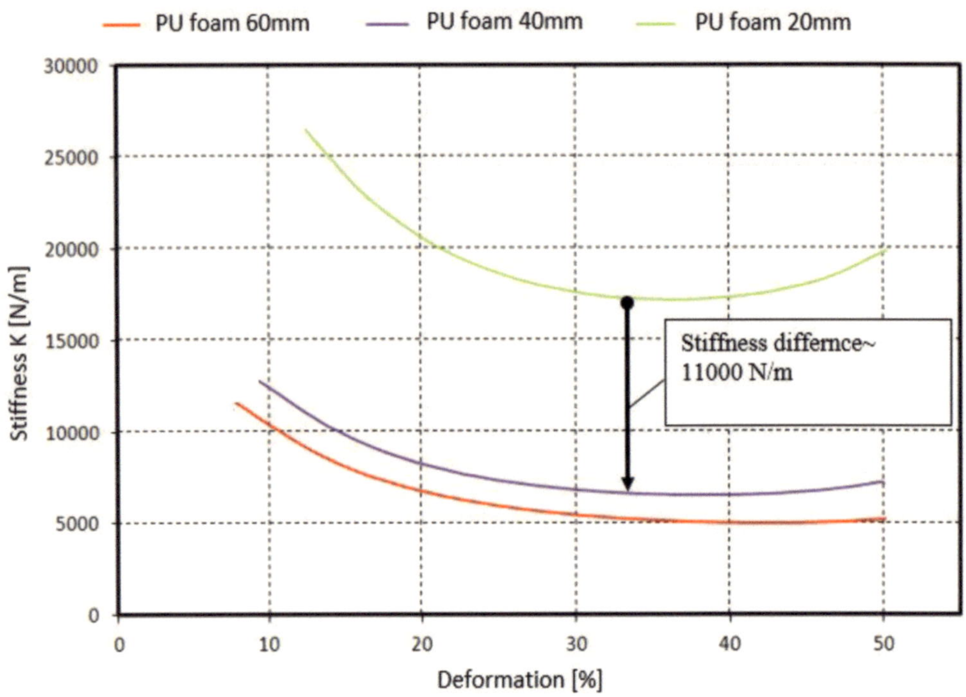

Figure 6. Dependence of stiffness on the deformation of PU foam samples with dimensions 100 × 100 × 60, 40 and 20 mm.

described by Eq. (23) is proportional to the area of hysteresis curve that describes the relation between tension and the relative deformation during one cycle of harmonic stress. Generally, with viscoelastic structures, it is true that in the harmonic excitation, the structure stress $\sigma(t)$ and deformation $\varepsilon(t)$ change in time, while $\varepsilon(t)$ has certain phase delay to the applied stress $\sigma(t)$, which is defined by Eqs. (4) and (5). Phase shift $\phi(t)$ between the tension and relative deformation lies during the harmonic excitation in the interval $\phi(t) \in (0, \pi/2)$.

$$\sigma(t) = \sigma \cdot \cos(\omega \cdot t + \phi) = \sigma \cdot \cos\phi \cdot \cos(\omega \cdot t) + \sigma \cdot \sin\phi \cdot \cos(\omega \cdot t + \pi/2) \tag{4}$$

$$\varepsilon(t) = \varepsilon \cdot \cos(\omega \cdot t), \tag{5}$$

Eq. (4) describing time dependency of the tension during harmonic compression can be further described in Eq. (6) expressing components of the dynamic module of the material structure.

$$\sigma(t) = E_P^{'} \cdot \varepsilon \cdot \cos(\omega \cdot t) + E_P^{''} \cdot \varepsilon \cdot \cos(\omega \cdot t + \pi/2), \tag{6}$$

where $E_P^{'}$ is a real component of the dynamic flexibility module describing durability properties of the material, and $E_P^{''}$ is imaginary component of the dynamic flexibility module describing dissipation of energy (loss module). Both modules are described by Eqs. (7) and (8), from which it is possible to obtain complex dynamic module E_P^{D} according to Eq. (9).

$$E_p^{'} = \frac{\sigma_0}{\varepsilon_0} \cdot \cos\phi, \tag{7}$$

$$E_p^{''} = \frac{\sigma_0}{\varepsilon_0} \cdot \sin\phi, \tag{8}$$

$$E_p^{'} = \frac{\sigma_0}{\varepsilon_0} \cdot \cos\phi, \tag{9}$$

where E_p^D is a complex dynamic module and i represents imaginary component.

To obtain mechanical properties during dynamic compression of the selected PU foams, measurements with 100 × 100 × 40 mm samples were conducted. All observed properties from these measurements can be summarized in these points:

- Determining the mechanical properties of selected samples with dynamic compression against a rigid plate without the initial deformation.
- Determining the mechanical properties of selected samples with dynamic compression against a rigid plate with the initial deformation.

3.3. Determining the mechanical properties of selected samples with dynamic compression against a rigid plate without the initial deformation

The experiment took place in the hydrodynamic laboratory (HDL). The measuring device was comprised of a hydraulic cylinder with an attached contraption to insert the sample. The contraption consists of two vertical supporting tubes put on the circular plate that were connected by a crosspiece from the top. In the middle of the crosspiece, an immovable tube pole was attached, with a 0.5 kN sensor placed on it. The sample was put between the upper and lower rigid plates. The arrangement of the conducted experiment is described in **Figure 7**. Input excitation harmonic signal (stationary periodical) was defined by Eq. (10). This signal is suitable for more than a study and experimental comparison of the material samples of different structures, because it is a base signal for the comparison and optimization of competent car seats [6].

$$y_{(z)} = A_{(z)} \cdot \sin(\omega t), \tag{10}$$

where $y_{(z)}$ is a defined cylindrical lift, $A_{(z)}$ is the input amplitude, and $\omega = 2\pi f$ is an angular velocity.

There are two possible approaches to define the input excitation of the hydraulic cylinder:

1. **Measurement with frequencies comprised in one input file** (one measurement with gradual change of the frequency value)
2. **Measurement with initial constant frequency value** (gradual measurement)

Measurements of the selected samples were conducted according to the method number 2—*measurement with a constant initial frequency value*. In total, seven measurements with a gradually rising frequency f were conducted, starting from 0.5, 1, 2, 3, 4, 5, and 8 Hz with constant

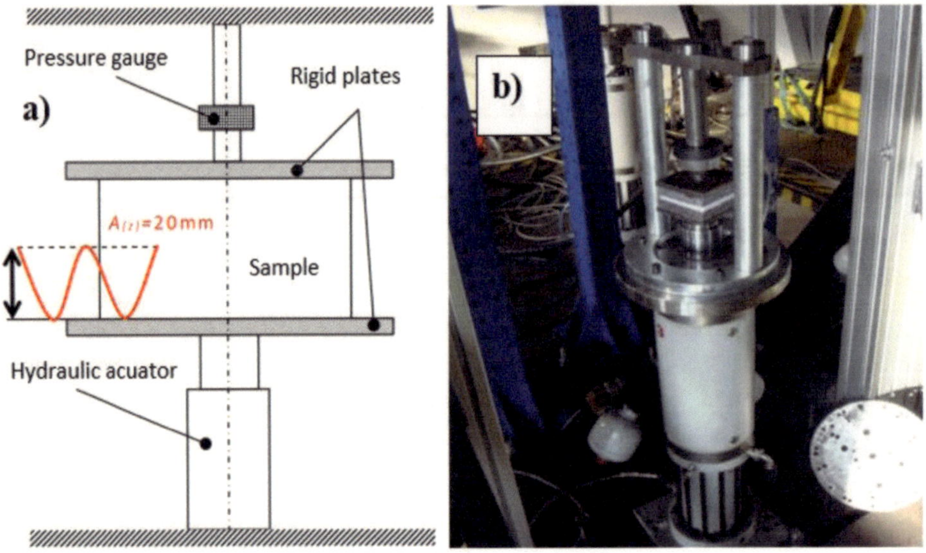

Figure 7. Determination of mechanical properties of samples of polyurethane foams during dynamic compression: (a) scheme, (b) realization of the measurement.

value of the amplitude $A_{(z)} = 20\ mm$ (i.e. up to 50% deformation) for the evaluation of five consecutive cycles. The difference of the harmonic course of a given input frequency is described in **Figure 8**, describing how the rising frequency value also raises the value of a phase shift (oscillation) of the hydraulic cylinder. Measurements were repeated three times. The resultant courses of the tested PU foam sample for individual frequency values during the fifth cycle are shown in **Figure 9**. The resultant courses of the dependence between pressure forces applied on the PU foam samples that were compressed against a rigid plate without the initial deformation for selected frequency values, which were 0.5, 2, and 4 Hz, are presented in **Figure 11**. The order of experiment where initial deformation is applied is shown in **Figure 10**.

The results of harmonic compression without the initial deformation in the fifth cycle of the PU foam sample (**Figure 9**) show that the change in frequency changes hysteresis force dependence on the pressure and relief. Looking at the PU foam sample, it is apparent that increasing frequency value 0.5, 2, and 4 Hz also increases the value of the force necessary to compress the material, but on the other hand starts dropping for frequencies 5 and 8 Hz. Maximum force value necessary to compress the PU foam sample was 191 N during 4 Hz frequency, while frequency 5 Hz slightly decreases the force value to 182 N and frequency 8 Hz requires only 165 N. This shows that the cell structure of the PU foam sample changes mechanical properties with the speed of deformation and change in frequency. It is apparent from the results that the PU foam sample changes its properties based on the speed of deformation, while the main influence can be seen in the presence of air in the foam structure. The air is not capable of getting back into the structure during unloading after reaching a certain strain rate; therefore, its influence is not as significant and the force value also decreases during compression.

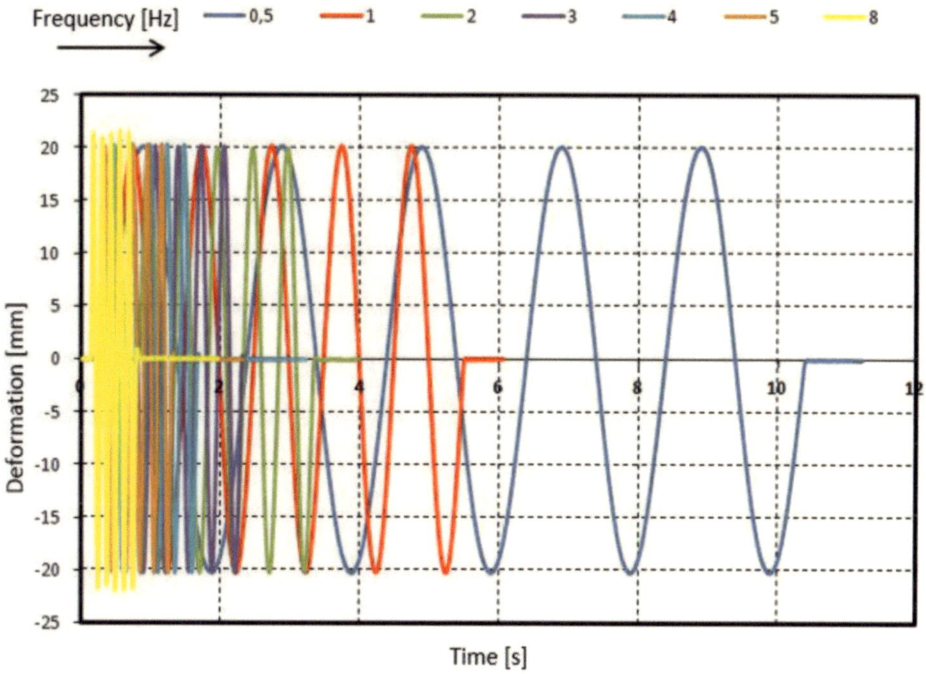

Figure 8. Input harmonic signals for measuring of dynamically compressed samples without initial deformation.

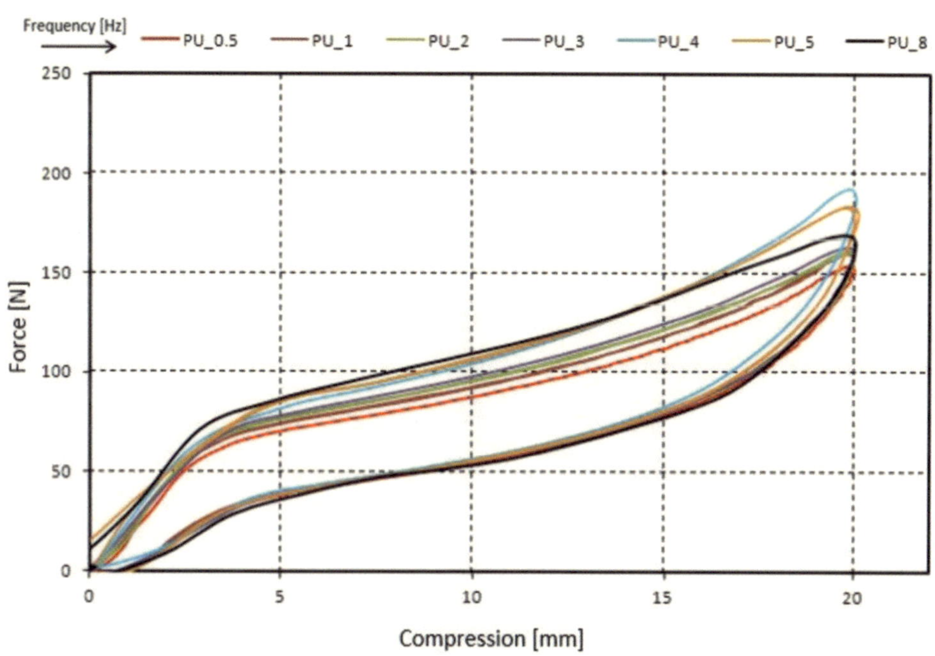

Figure 9. Dependence of force on deformation of dynamically compressed samples without initial deformation.

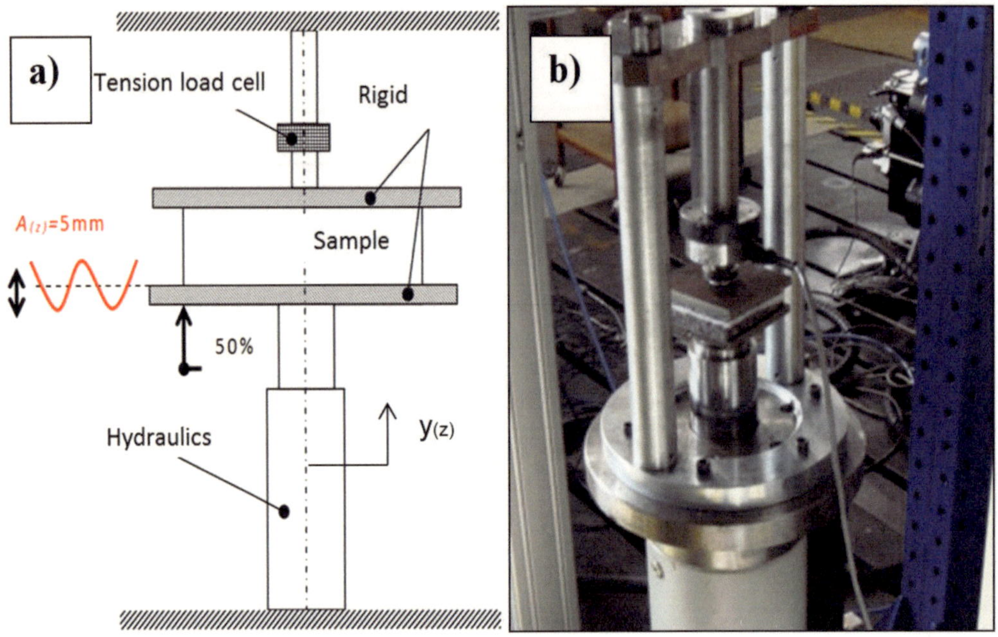

Figure 10. Determination of mechanical properties of samples of polyurethane foams during dynamic compression with initial deformation: (a) scheme, (b) realization of the measurement.

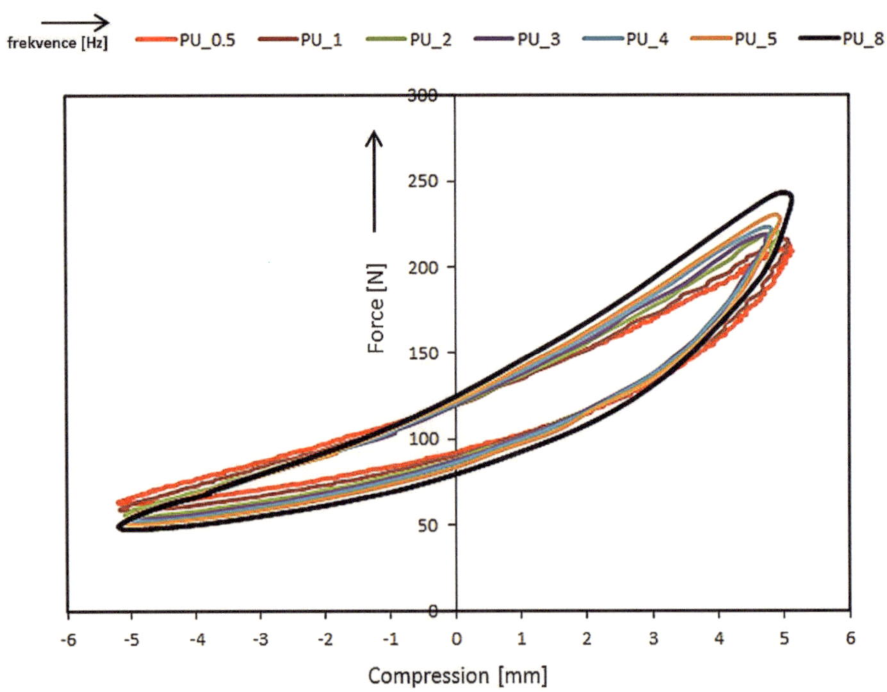

Figure 11. Dependence of force on deformation of dynamically compressed samples with initial deformation.

3.4. Determining the mechanical properties of selected samples with dynamic compression against a rigid plate with the initial deformation

The experiment with chosen samples was conducted 60 min after the first measurement, a long enough time necessary to relax the tested sample. The only difference in measurement was that the tested sample was compressed by the upper rigid plate by 20 mm to its initial 50% deformation. After the initial compression, seven measurements with gradually rising frequency f were conducted, from 0.5, 1, 2, 3, 4, 5, and 8 Hz but with the amplitude at $A_{(z)} = 5\ mm$ for five repeating cycles. The arrangement of the experiment is shown in **Figure 10**. The resultant courses of the tested PU foam sample for individual frequency values during the fifth cycle are shown in **Figure 9**.

Results for the PU foam sample with initial 50% deformation in the fifth cycle of the repeated compression (**Figure 11**) show that the change in frequency also increases the force necessary to compress the material that reaches maximum 246 N and frequency of 8 Hz. This is different, compared to the dynamic measurement of the sample without the initial deformation, because beginning with 5 Hz value, the necessary compression force began to drop. The courses also have a very similar character while comparing samples with different frequency of 0.5, 2, 4, and 8 Hz, which creates so-called "banana curve" shown in **Figure 11**. While comparing the courses during 8 Hz frequency, a dynamic ratio i_d expressing the ratio between the maximum force value and minimum force value of the force relief during compression in the PU foam samples is $i_d = 4.86$ (min 50 N and max 243 N). The higher the value of the dynamic ratio between maximum and minimum force, the faster the material recovers, because there is a greater energy return to recover the material. From the maximum force value necessary to compress the tested sample during the dynamic measurement, it is apparent that the value is higher than during a static compression; therefore, the rigidity of the sample increases. It is possible to determine the dynamic flexibility module E_P^D is greater than static flexibility module E_P^S.

3.5. Measuring the relaxation of the chosen material samples

Also, it is important to compare the mechanical properties during a long-term compression. As it was stated, the cellular structure of the PU foam becomes more supple under constant pressure, and the increase in deformation grows $\varepsilon(t_2)|_{\sigma=konst} > \varepsilon(t_1)$, i.e. the structure "melts," or during the constant deformation, $\varepsilon = konst.$ relaxes and the tension gradually decreases $\sigma(t_2)|_{g=konst} < \sigma(t_1)$. This is true in general for all materials with viscoelastic properties. According to Refs. [1, 6], it is more advantageous to measure the relaxation of material for the evaluation of mechanical properties, because the "melting" of the structure during the constant compression is minimal and almost negligible (the significance grows during long-term measurements—weeks, months—in high temperatures). Comparison of the compressed samples was conducted to evaluate the relaxation of material. The experiment was conducted on the same device just as during the static testing (**Figure 4**). The relaxation properties were compared for the PU foam samples $100 \times 100 \times 40$ mm. The observed properties obtained from these measurements can be summarized in the following points:

- determining the relaxation of the chosen samples compressed by the rigid plate into the constant deformation value.
- determining the relaxation module for the chosen samples.

3.6. Determining the relaxation of the chosen samples compressed by the rigid plate into the constant deformation value

The sample was compressed by the rigid plate into the constant value atduring 10, 25, 50, and 65% deformation, while measuring the material response to the compression during 3600 s. Resultant courses of the sample force response dependence to time are shown in **Figure 12**.

The resultant comparison of the course of relaxation of the PU foam sample shows minimum relaxation during low deformation values (10 and 25%); on the other hand, there was an apparent decrease in force over time during 50–65% deformation. The PU foam sample had a decrease in force of 34 N during 65% deformation (initial force value was 118 N and the final force value was 84 N). The drop in force overtime can be converted to the decrease in tension over time, which can then express the values of the relaxation modulus *G(t)* according to Eq. (14) describing the decrease in tension in the material structure in a time sequence. Values *G(t)* for each deformation differed in the PU foam sample. This can be explained by the cellular structure with low initial deformation of 10% having greater rigidity and cellular structure with greater initial deformation having less air in its structure. In fact, the structure closes and the permeability decreases, and therefore the air cannot return to the structure. Relaxation module values for the tested samples are shown in **Table 2**.

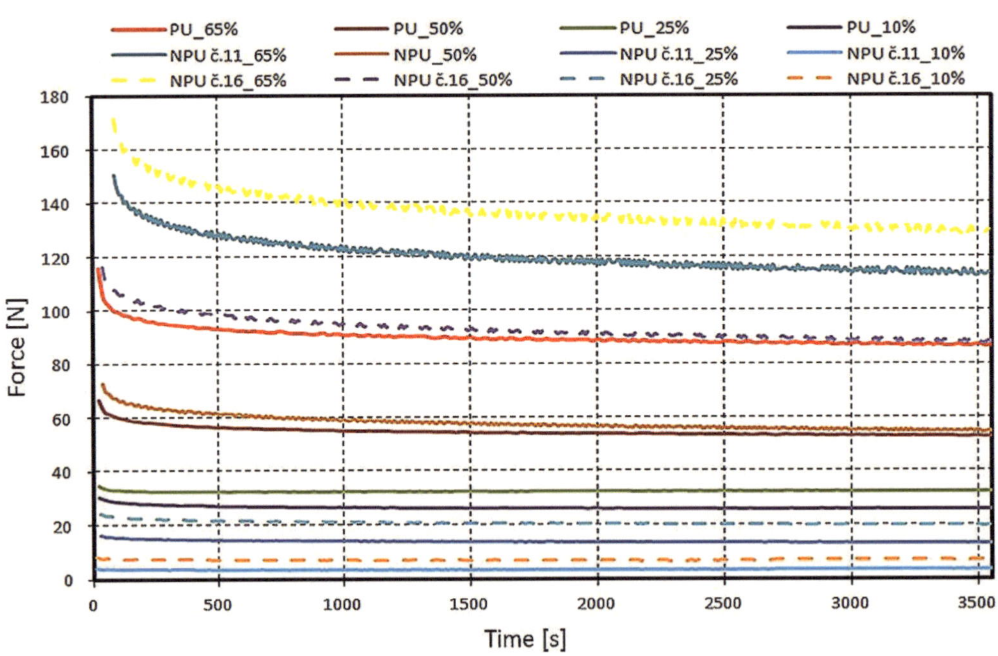

Figure 12. Comparison of relaxation experiment of PU foam.

Time t [s]	500	1000	1500	2000	2500	3600	Deformation [%]
G(t) [KPa] of PU foam	0.280	0.270	0.265	0.240	0.240	0.240	10
	0.132	0.128	0.126	0.126	0.126	0.126	25
	0.112	0.108	0.106	0.102	0.100	0.099	50
	0.143	0.137	0.134	0.131	0.128	0.128	65

Table 2. Results of relaxation modulus $G(t)$.

3.7. Mathematic-physical description of mechanical behavior of PU foam samples

Mechanical properties of the PU foam samples have a non-linear course during compression. This is a viscoelastic behavior with great, almost returnable, transformation, in the order of 92 ± 3% [12]. During the force relief on the foam, the hysteresis loop manifests, based on the volume weight, rigidity K, and damping coefficient η_t with gradual relaxation of the cell structure, that is more significant during the quasi-static compression than dynamic, among others [6, 12]. The inability of rapid recovery after the deformation also materializes. During the compression, the non-linear behavior of PU foam sample is characterized by three areas: first, initial solidification; second, steady course of deformation during the minimum gain in absolute tension; and third, final and significant solidification of the cell structure. Characteristic course of tension in the dependence on the transformation of the tested PU foam sample is shown in **Figure 13**.

The graphic course shown in **Figure 13** characterizes nonlinear dependence of tension on the transformation, gained during the experimental measurement of the PU foam sample number 3 (**Table 1**) compressed to 98 ± 2% transformation. The sample number 3 had characteristic **area 1** —lasting up to ~12 ± 3% transformation, which was given by the elastic, almost linear, course with a steep onset of the tension caused by initially fast solidification of the cellular structure, where the inclination of the curve is dependent on the strain rate. **Area 2** can be set for the range of 15–50 ± 5% transformation, where so-called plateau takes place (even temporary stabilization) meaning the absolute increase in tension depending on the transformation is minimal. In the **area 3**, approximately from 60 ± 7 up to 92 ± 3% deformation begins a steep exponential course caused by the final compression of the structure (the structure starts to almost mash). Already, in the year 1969, Rush [13] published analytical model closely describing the behavior of the compressed PU foam. The author based his assumptions on the constitutive equation describing the response of the material after compression, that is characterized by a constant module of flexibility E, the size of deformation transformation ε, and the compression function $\Lambda(\varepsilon)$, that can be described by Eq. (11), where E presents module of foam flexibility, ε is deformation, and $\Lambda(\varepsilon)$ is compression function.

$$\sigma = E \cdot \varepsilon \cdot \Lambda(\varepsilon), \tag{11}$$

Constitutive Eq. (11) was improved upon by Schwaber and Meinecke [14] in 1971 and by Nagy et al. [15] in 1974 by the functional dependence of variable flexibility modulus E on the strain rate $\dot{\varepsilon}$, which is described by Eqs. (12) and (13). These relations can be used to determine immediate rigidity in the structure, because the weight of the structure $m(\varepsilon, \dot{\varepsilon})$ and the flexibility

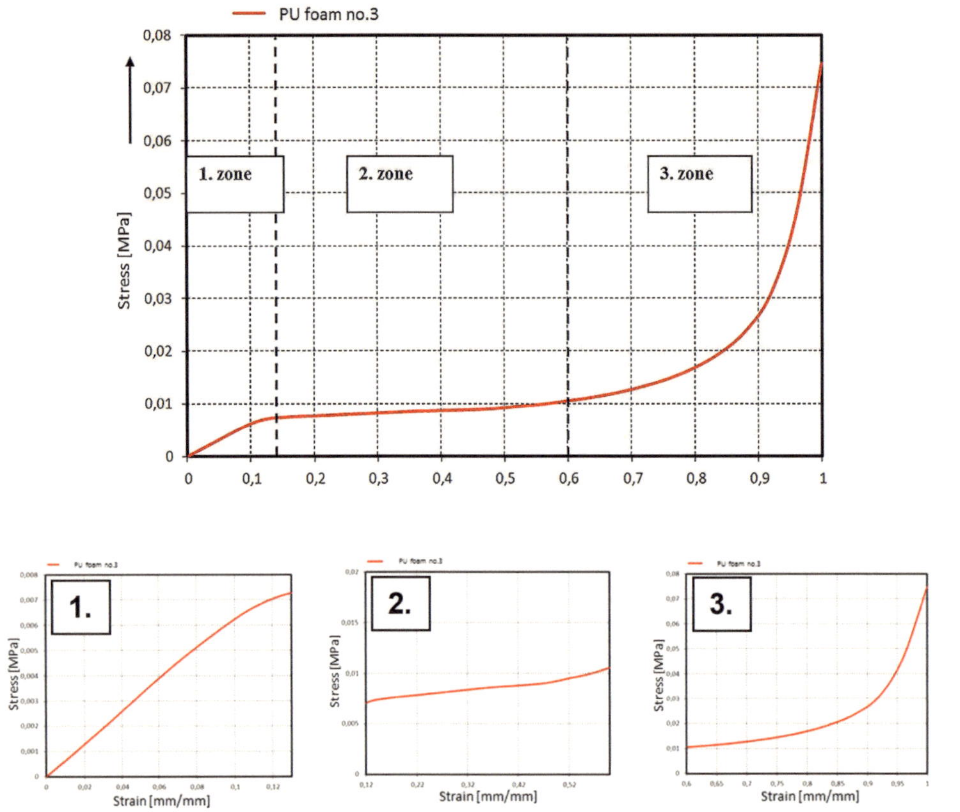

Figure 13. Nonlinear dependence of stress of PU foam sample (top), and the characteristic courses in three main areas (bottom).

modulus $E(\dot{\varepsilon})$ are variable and depend on the immediate state of the deformation or transformation ε.

$$\sigma = E(\dot{\varepsilon}) \cdot f(\varepsilon) \cdot \varepsilon, \tag{12}$$

$$\sigma = m(\varepsilon, \dot{\varepsilon}) \cdot f(\varepsilon), \tag{13}$$

where $f(\varepsilon)$ describes polynomial function describing the course during compression, $m(\varepsilon, \dot{\varepsilon})$ is the weight of the structure that is variable on deformation ε and the strain rate $\dot{\varepsilon}$.

Characteristic properties of the PU foam sample during compression are influenced mostly by the size of the deformation $\varepsilon(t)$ but also time t that takes to compress the cellular structure. Related to this is a physical effect called creep (thinning the structure); in other words, the structure of the PU foam becomes suppler under constant pressure and the deformation increases $\varepsilon(t_2)|_{\sigma=konst} > \varepsilon(t_1)$. Similarly behaves the structure during relaxation of the material—during constant deformation $\varepsilon = konst$ (for example, repeated cyclic compression up to 50% transformation and the material fatigue), the tension gradually decreases. It is possible to describe this physical effect by the value of the relaxation modulus $G(t)$ (14) described by the relation between acting tension $\sigma(t)$ and a constant $G(t)$ deformation $\varepsilon = konst$. In Ref. [1], it is stated that increasing deformation of the PU foam sample decreases the value $G(t)$ where a setting of relaxation modulus values materializes during long-term test for values of small and large deformations as stated in **Figure 14**. This effect is caused by viscoelasticity of the material (this does not have to be true in all viscoelastic materials, and it even can be reversed).

$$G(t) = \frac{\sigma(t)}{\varepsilon_{konst}} \tag{14}$$

Mechanical properties of the PU foam are further determined by the temperature T. Authors in [16] already included the influence of temperature T in the analytical model and consecutive constitutive relation is further described by Eq. (15). The relation character (15) was further improved in Ref. [17] by material constants a, b reflecting even the strain rate $\dot{\varepsilon}$ in relation to morphology of the foam, described by Eq. (16). Authors already establish constants in the model, which statistically describe the frequency of air bubbles in the structure. It is necessary to mention that the significant influence of temperature on the foam behavior happens according to the experience of the producers only in foams used to fill the comfort layers of

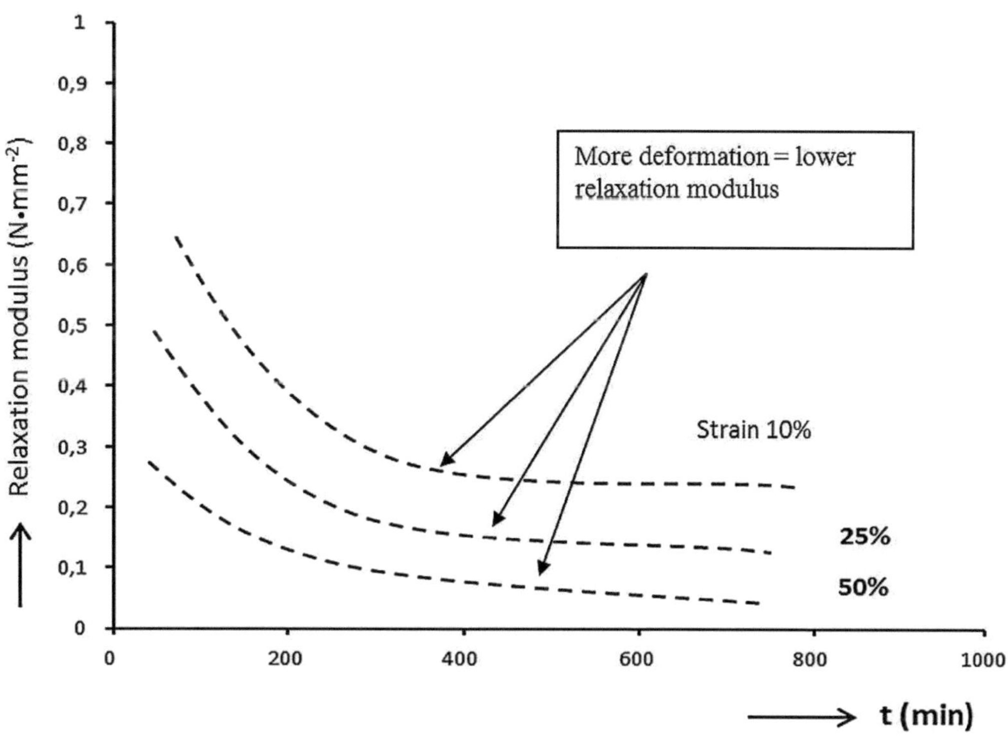

Figure 14. Relaxation modulus of viscoelastic material depending on time.

the car seats during high deviations from the room temperature (for example extreme cold −40°C or on the other hand extremely high temperatures above 70°C).

$$\sigma = h(T, \dot{\varepsilon}) \cdot g(\rho) \cdot f(\varepsilon), \tag{15}$$

$$\sigma = \sigma_0(\varepsilon) \cdot h(T) \cdot \left[\frac{\dot{\varepsilon}}{\dot{\varepsilon}_0}\right]^{a+b\tau} \tag{16}$$

where $h(T, \dot{\varepsilon})$ expresses a step function related to the temperature T and the change in the strain rate $\dot{\varepsilon}$, $g(\rho)$ is experimentally set value related to the structure density, and a, b are material constants $(a, b \geq 1)$.

Viscoelastic behavior of the PU foam can be significantly described using rheological models, i.e. Refs. [8, 14, 16, 17]. Rheology is a science investigating mostly changes in tension σ and transformations in relation to time t and the strain rate $\dot{\varepsilon}$. It stems from the transformation of continuum and therefore does not investigate the structure mechanics (structure morphology and typology, number of air bubbles) compared to the previous relations. Models are created via system of connections of various combinations of Hook elastic elements (springs) and Newton viscose components (damper). They allow for approximate description of non-linear behavior of materials structures including PU foam, by linear components in a various number and combination. It is possible to also study the relaxation and material creep from the obtained relations from the rheological models. The advantage of rheological models is mostly independence on material models in the finite element method programs. In the wide range of publications concerning modeling of PU foam, i.e. Refs. [6, 8], a one-dimensional Kelvin or Maxwell rheological model was used. Authors often mention good congruence of the resultant dependencies of the rheological model in comparison with experimental measurements. Complex cellular structure of the PU foam causes more complicated rheological behavior, where deformation always contains part of elastic, viscose, or sometimes permanent deformation. From the refining values of these one-dimensional rheological models, we can consecutively obtain their n-parameter expansion, shown in **Figure 15**.

It is possible to create rheological model, which will be getting significantly (in limit) close to the experimentally measured data, by a mutual composition and combination of n-parametric Kelvin and N-parametric Maxwell model, better said by a various number of compounded Hook and Newton elements. According to such model, it will be possible to create a corresponding mathematical expression describing mechanical properties of the PU foam sample, especially the dependence of force/compression or tension/transformation or also dampening and elastic properties (rigidity coefficient K and a dampening coefficient η_t on the immediate transformation). This can be achieved by a rheological model according to **Figure 16**. This is a modified n-parametric Tucket model. This model shows practically three characteristic areas. The first part is comprised of a sum of m-number of springs (while $m < n$), which describes initial solidification characteristic by an elastic deformation. The second part represents an n-parametric Kelvin model, comprised by a sum of m—parallel-connected springs and dampers, which describe the delayed viscoelastic deformation. The third part is then characterized by a sum of m—number of dampers (while $m < n$), describing final remaining deformation after tension ceases to be applied. The advantage of this model is

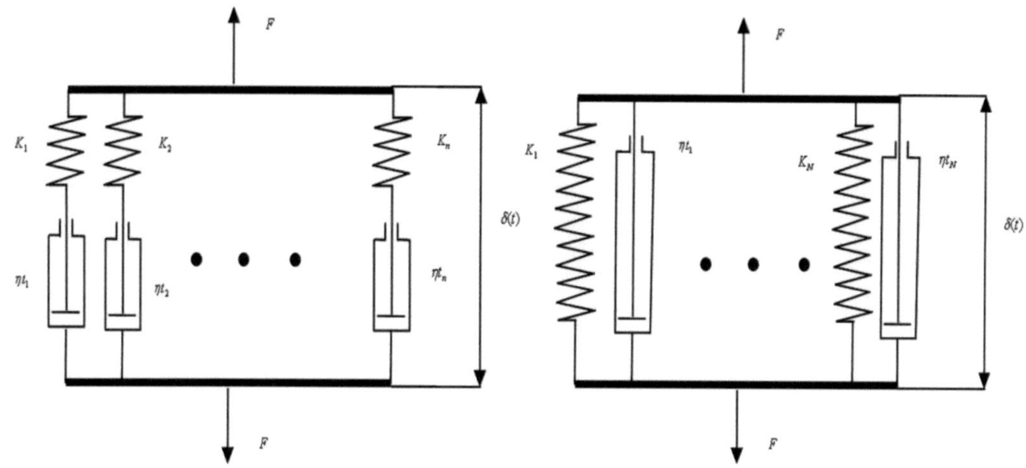

Figure 15. *N*-parametric rheological models: Maxwell model (left) and Kelvin model (right).

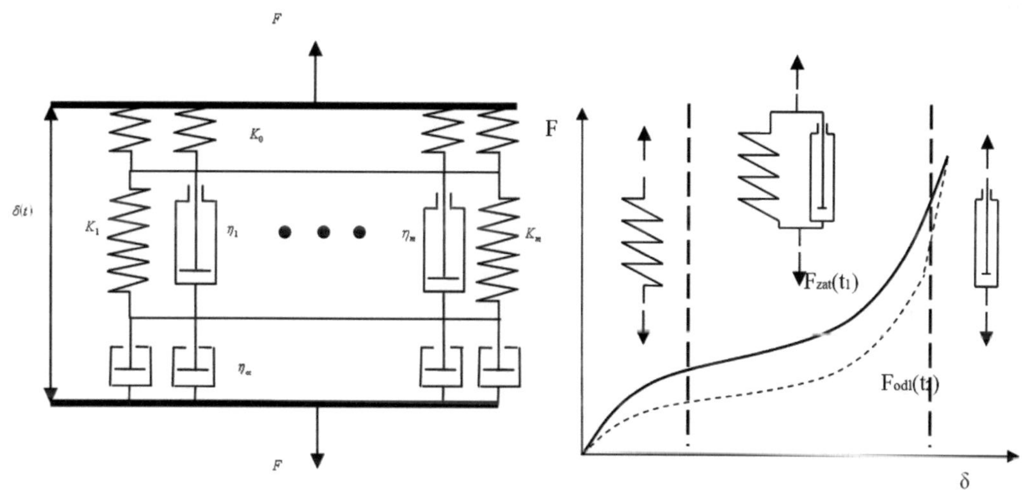

Figure 16. *N*-parametric Tucket model allowing description of nonlinear behavior of PU foam.

mainly the fact that the faster the deformation is supposed to happen, the greater the inhibition effect of the viscoelastic element; therefore, greater force must be applied to achieve desired deformation. The model therefore describes the behavior of viscoelastic material that increases resistance to compression of the applied force by an inner viscose medium. After the force ceases to be applied (final value of compression $\delta|_{\tau=\max}$), the deformation stays maintained in a limit moment $\varepsilon(t) \equiv \delta|_{t=0}$, and after some time, recovery follows.

According to the following rheological model (**Figure 16**), we can construct a corresponding mathematical expression of the compressed PU foam sample by the Eqs. (17)–(24) describing the dependence of the force on compression, for example, rigidity, dampening, or creep suppleness.

$$\sum_{i=1}^{m}\left(K_{PU}(\tau_1)\cdot\bar{\delta}(\tau_1)+\frac{d}{d\tau_1}(\eta_{tPU}(\tau_1)\bar{\delta}(\tau_1))\right)_t = F_z(\tau_1)\ \ pro\ \ m < n, \tau_1 < t, \tag{17}$$

$$\sum_{i=1}^{m}\left(K_{PU}(\tau_2)\cdot\bar{\delta}(\tau_2)+\frac{d}{d\tau_2}(\eta_{tPU}(\tau_2)\bar{\delta}(\tau_2))\right)_t = F_o(\tau_2)\ \ pro\ \ m < n, \tau_2 < t \tag{18}$$

$$\bar{\delta}(\tau_1) = \delta(\tau_1) - \frac{F_z(\tau_1)}{K_{oPU}} - \frac{1}{\eta_{tPU}}\int_0^{\tau_1} F_z(\tau_1)dt, \tag{19}$$

$$\bar{\delta}(\tau_2) = \delta(\tau_2) - \frac{F_o(\tau_2)}{K_{oPU}} - \frac{1}{\eta_{tPU}}\int_0^{\tau_2} F_o(\tau_2)dt, \tag{20}$$

where $F_z(\tau_1)$ describes stress force in time $\tau_1 < t$, $F_o(\tau_2)$ describes relieving force in time $\tau_2 < t$, $\bar{\delta}(\tau_1)$ describes the duration of the material compression, that is different in time during hysteresis (significantly longer, shorter, or negligible), among others due to resistance of the material compared with the relief time $\bar{\delta}(\tau_2)$, $K_{PU}(\tau_{1,2})$ is the immediate value of rigidity of the PU foam sample during compression, and stress relief $\eta_{tPU}(\tau_{1,2})$ is the immediate value of PU foam sample dampening during compression and stress relief.

The functional dependence of the total rigidity and total dampening of the structure can be consequently described.

$$K_{PU}(t) = \sum_{i=1}^{n}\left(\frac{\frac{\bar{\delta}(\tau_2)}{\tau_2}\cdot F_z + \frac{\bar{\delta}(\tau_1)}{\tau_1}\cdot F_o}{\left(\frac{\bar{\delta}(\tau_1)}{\tau_1}+\frac{\bar{\delta}(\tau_2)}{\tau_2}\right)\cdot\bar{\delta}(t)}\right)_t, \tag{21}$$

$$\eta_{tPU}(t) = \sum_{i=1}^{n}\left(\frac{\frac{\bar{\delta}(\tau_1)}{\tau_1}}{\left(\frac{\bar{\delta}(\tau_1)}{\tau_1}+\frac{\bar{\delta}(\tau_2)}{\tau_2}\right)\cdot\bar{\delta}(t)}\int_0^t (F_z(\tau_1) - F_o(\tau_2))dt\right)_t, \tag{22}$$

where $K_{PU}(t)$ describes total rigidity of the PU foam sample, and $\eta_{tPU}(t)$ describes total dampening of the sample.

Also through the work difference, it is possible to obtain the relation for the dissipated energy $\vartheta(t, \delta, T)$, which describes energy that can be absorbed by the material.

$$\vartheta(t, \delta, T) = \sum_{i=1}^{n}(W_z - W_o)_t = \sum_{i=1}^{n}\oint(F_z - F_o)_t dl, \tag{23}$$

where W_z, W_{od} describe work that the material does during compression and unloading.

Using this model, it is possible to describe creep suppleness $\Theta(t)$ in Eq. (24).

$$\Theta(t) = \frac{1}{E_0}(1 - e^{\frac{t}{\tau}}) + \sum_{i=1}^{m} \Theta(m) \cdot \left(1 - e^{\left(-\frac{t}{\tau}\right)^m}\right) \quad for\ m < n, \tau_2 < t, \tag{24}$$

where E_0 is the initial stiffness module.

Results of the n-parametric Tucket model expressed as the dependence of the tension on the transformation are shown in **Figure 17** where courses are in a very good agreement with the experiment. Correlation coefficient comparing between surveyed courses has a value of ~0.978. Certain difference is probably given by a fact that the rheological model does not include the structure morphology, i.e., the cell walls bend and from a certain phase of compression create friction between one another. According to Eqs. (21) and (22), it is possible to determine the course of rigidity and dampening of the PU foam sample (**Figure 18**).

Using appropriate relations, we can determine absolute value of deformation energy $E(\mu)$ or deformation work $W(\mu)$ that is necessary to apply to compress such structures. Fibrous structures are not conservative, therefore during the compression depending on the character of the deformation from the initial to final state, according to Ref. [18], it can be considered that

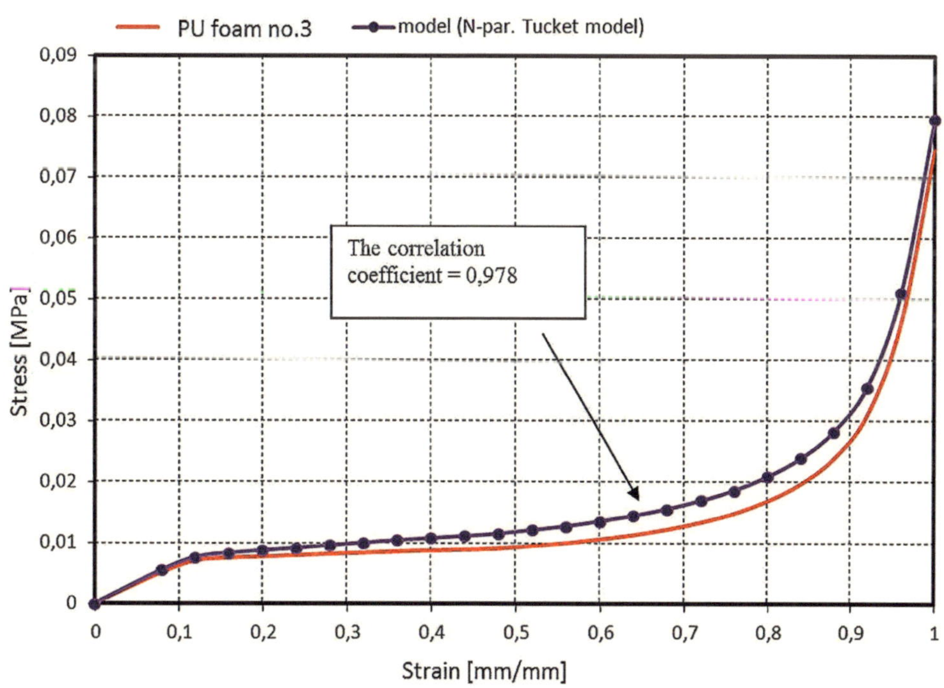

Figure 17. Comparison of nonlinear dependence of stress on strain of PU foam, experiment (line), n-Tucket n-parametric model (dotted line), and correlation coefficient 0.978.

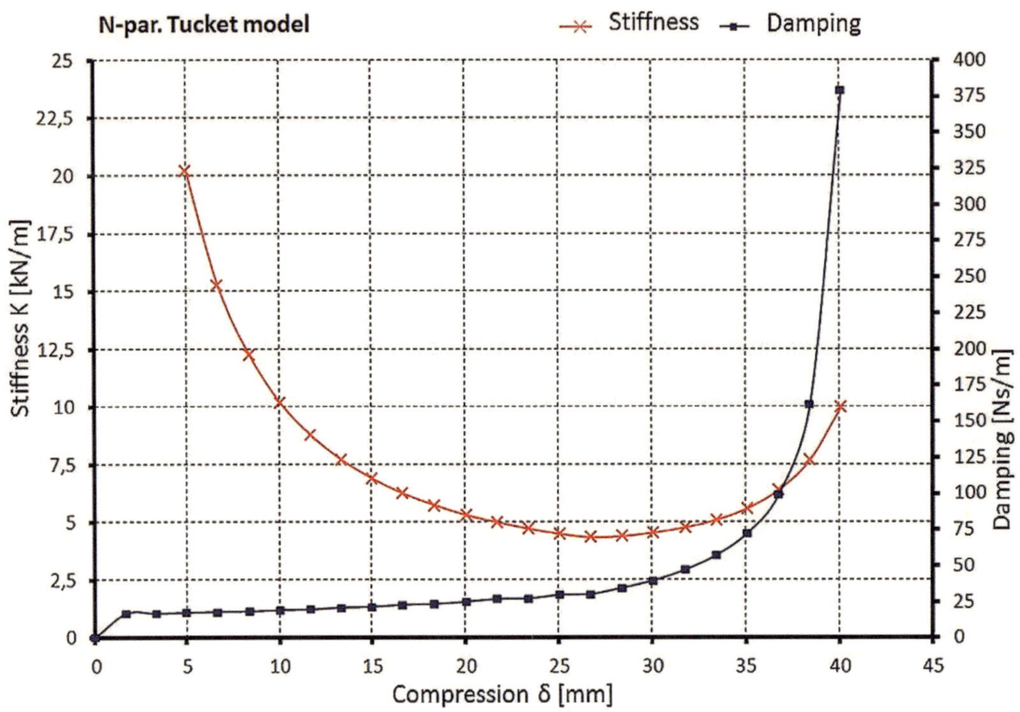

Figure 18. *N*-parametric Tucket model: course of stiffness and damping depending on compression.

elementary work growth dW used to compress structure is directly proportional to the elementary gain of the deformation energy dE according to the Eq. (25) where elementary energy gain is a total function differential $E(\mu) = E_\mu(\varepsilon_i)$, which can be described by Einstein summary convention according to Eq. (26). Deformation into individual directions ε_i is possible to further describe using the filling μ according to Eq. (25). It is then possible to gain derivation change (transformation) of the filling in dependence to the deformation ε_i as per Eq. (26).

$$dW(\mu) = CdE(\mu) \ \ pro \ \ C\geq 1, \tag{25}$$

$$dE = \frac{\partial E}{\partial \varepsilon_i} d\varepsilon_i \ \ pro \ \ i = 1, \ldots, 3, \tag{26}$$

$$\mu = \frac{V_V}{V_C} = \frac{V_V}{(1+\varepsilon_i)(1+\varepsilon_j)(1+\varepsilon_k)} = \frac{\mu_0}{(1+\varepsilon_i)(1+\varepsilon_j)(1+\varepsilon_k)} \tag{27}$$

$$\frac{\partial \mu}{\partial \varepsilon_i} = \frac{\partial}{\partial \varepsilon_i}\left[\frac{\mu_0}{(1+\varepsilon_i)(1+\varepsilon_j)(1+\varepsilon_k)}\right] = \frac{-\mu_0}{(1+\varepsilon_i)^2(1+\varepsilon_j)(1+\varepsilon_k)}, \tag{28}$$

where $W(\mu)$ is a deformation work ($W(\mu) = \int \sigma d\varepsilon$), C is a constant of proportionality, $\varepsilon_{i,j,k}$ describes deformation into the main direction of extension, and μ_0 is the initial filling, i.e. $\mu_0 = V_V$.

The structure creates resistance during deformation against the compression described by the distribution of Cauchy (real) stress tensor σ_{ii} related to the areas in the deformed continuum,

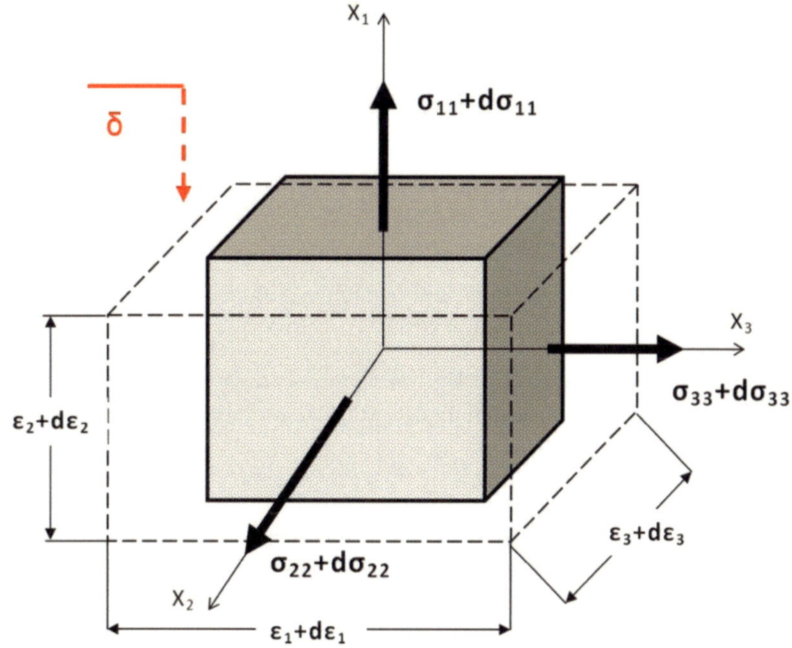

Figure 19. Elementary continuum of compressed structure.

which is well described in Ref. [19]. This is described in **Figure 19**, which shows base configuration of elementary continuum cube, where $\varepsilon_i = 0|_{t=t_0}$ transforms during increases in deformation into the deformed shape $\varepsilon_i \neq 0|_{t>t_0}$. Using Cauchy stress tensor σ_{ii}, it is possible to describe Eq. (25) by the sum of contributions to the main directions of deformation, which is given by Eq. (29). Total stress σ_{HMH} described by the von Mises hypothesis (Eq. (30)) based on normal parts of Cauchy stress tensor σ_{ii} simply describes the compression pressure p_k.

$$\sum_{i=1}^{3} \sigma_{ii} d\varepsilon_i = C \sum_{i=1}^{3} \frac{\partial L}{\partial \varepsilon_i} d\varepsilon_i \tag{29}$$

$$\sigma_{HMH} = \sqrt{\frac{1}{2}\left[(\sigma_{11} - \sigma_{22})^2 + (\sigma_{22} - \sigma_{33})^2 + (\sigma_{11} - \sigma_{33})^2\right]} \tag{30}$$

where σ_{HMH} is a total (reduced) tension according to the hypothesis HMH (Huber, von Mises, Hencky), $\sigma_{11}, \sigma_{22}, \sigma_{33}$ are the Cauchy stresses in the main directions of the basic coordinate system X_i.

3.8. Summary of the property analysis of selected PU foam samples

The property analysis of the chosen PU foam construction samples determined that the tested samples have a low-permeable shell caused by the filled structure in form, and inner structure is permeable. Depending on the specific weight ρ, which can move from 47 to 51 kg

m^{-3},the packing density Ψ (1) increases while influencing the average cell size of the material structure (**Table 1**). Cellular structure of the PU foam sample influences strain-stress curve during compression. Typical course is characterized by the initial rigidity depending on the strain rate, given stable area so-called plateau and the final steep exponential increase in force.

Mechanical properties are significantly influenced by the character of the cellular structure, and they can be summarized as follows:

- Deformation ε is the function of not only tension σ, but also time t, and in the examined area (up to 95 $\pm$ 3% transformation), it can be considered **reversible**.
- Material deformation is **dampened** by inner viscose resistances, i.e., dampening of the cellular structure, η_{tPU}, therefore cannot be realized immediately.
- The faster the deformation happens, the more intense the dampening effect of the material viscosity materializes, but also the air viscosity, which cannot be squeezed out of the cellular structure immediately and, therefore, manifests as an initial significant increase in rigidity (the slower the compression, the lower the initial rigidity $\rightarrow$ the initial rigidity is nearing the behavior of the so-called plateau).
- **Recovery** manifests (recovery—the ability to immediately recover after deformation) given viscoelastic properties determined by hysteresis.
- **Relaxation** manifests, i.e., decrease in tension in prestressed condition $\sigma(t)|_{z-konst} < \sigma(0)$
- **Creep** manifests, i.e., the growth in deformation under constant pressure $\varepsilon(t)|_{\sigma-konst} > \varepsilon(0)$
- Mathematic-physical description of PU foam mechanical properties can be described by a constitutive relations and by **rheologic models**, for example, modified n-parametric Tucket model according to which it is possible to express the coefficient of rigidity and dampening of PU foam.
- For a qualitative analysis of the PU foam during compression, or for wider knowledge of mechanical properties that cannot be sufficiently measured nor mathematically described (distribution of main tensions and transformation in individual directions, contact pressures), it is suitable to construct **model simulation** of mechanical properties in the setting of the finite element method.

4. Model simulations of mechanical properties of selected samples

The analysis and measurements of mechanical properties of PU foam samples are generally limited only to certain information, and therefore, it cannot tell us the immediate distribution of deformation and tension in the material structure. This is because the options for measurements are limited; options to place sensors and some properties cannot be well measured (for example, distribution of main tension and deformation of the cellular structure). In this case, a compilation of appropriate model simulation according to the numerical method becomes a viable option. Programming a model simulation in FEM setting is the most significant to our purpose, but certain ways also offer other numerical methods, for example, method of discrete elements (MDE), method of boundary elements (BEM), or method of finite volume (FVM). The FEM method was used exclusively in this work. Mechanical compression of the selected PU foam samples creates many heterogeneous attributes in its inner structure, which change with the size of deformation, as was stated. While modeling such structures, Refs. [8, 20] agree that it is necessary to simplify or ignore certain characteristic attributes, and both state that a great problem of modeling non-linear attributes is mostly the description of the main tensions in short time differentiations $\Delta t = t_{i+1} - t_i$. The solution of peripheral problem of great deformations created by compression of the sample further lies not only in the correct input of peripheral conditions and material properties, but also in the construction of proposed net of finite elements. FEM programs are currently very well worked out and allow for the conversion of continuous problem solution to the final solution, where it is possible to suggest appropriate geometrically simple partial sub-regions (finite elements) for approximate solution in the preprocessor.

Let $\aleph \subset R^3$ is a continuous area of three-dimensional space. Their boundaries will be Γ, where Γ is the so-called Lipschitz boundary, and let an approximation of the selected basis functions are derived over each finite element with dimension l, because any continuous function may be represented by linear combinations of algebraic polynomials converging to a continuous solution i.e. $\lim_{l \to 0} \to \xi \approx 1$. Thus, FEM can be understood as a special type of a variation method using mathematical description of solved problem. Current substantial commercial FEM software (for example, Ansys, Abaqus, Permas, LS-Dyna, Marc, and PAM CRASH) allow to assemble and then solve the loading of nonlinear materials not only with viscoelastic properties using mathematical relationships based on continuum mechanics and rheological model (for example, Kelvin model, Maxwell model, and so on). Also they allow with sufficient accuracy the studying and modeling of qualitatively more complex problems such as contacts of two or more bodies (i.e., the interaction between the material and probe or human body).

4.1. Selection of suitable software for the assembly of FEM model

In this study for all model simulations, software PAM CRASH was selected. It is the FEM software from ESI-Group company (http://www.esi-group.com/) that is used for the study of nonlinear isotropic and anisotropic materials and contact problems in quasi-static and dynamic states. Similar to FEM software as LS-Dyna, Abaqus Explicit, and ANSYS Explicit. The basic principle of explicit methods is second Newton's law, which may be expressed in a matrix form by Eq. (31).

$$M \cdot \ddot{u} = F^E - F^I, \tag{31}$$

where M is a matrix of mass, $\ddot{u}$ is the acceleration matrix of node vectors, F^E is the matrix of the vector of external forces acting on the node, and F^I is a matrix of vectors of internal forces (volume forces).

The matrix of acceleration vectors, where the acceleration expresses second derivation of the searched (unknowns) displacements (Eq. (32)), can be obtained according to Eq. (31). Then, vector matrix of internal and external forces can be expressed by Eqs. (33) and (34).

$$\ddot{u} = M^{-1} \cdot F^E - F^I \tag{32}$$

$$F^I = \sum_{\varepsilon=1}^{N_\varepsilon} \left(\int_{\Omega} B^T \cdot \sigma_n d\Omega + F^{Hurg} + F^{kont} \right), \tag{33}$$

$$F^E = \sum_{\varepsilon=1}^{N_\varepsilon} \left(\int_{v_\varepsilon} \rho \cdot \kappa_{i\infty}^{\varepsilon} dv_\varepsilon + \int_{s_\varepsilon} \chi_i \cdot \vartheta_{\infty}^{\varepsilon} ds_\varepsilon \right) \tag{34}$$

where B is matrix of basic functions of the strain, F^{kont} is the vector of the contact forces, F^{Hurg} is a vector of hourglassing damping forces, σ_n is the matrix of acting stress in member, ρ is density of the member, κ_i is vector of volume forces, and χ_i is vector of surface forces.

Founded matrix of displacement vectors u (Eq. (36)) can be expressed by an integration of the acceleration $\ddot{u}$ or velocity of displacements $\dot{u}$ (Eq. (35)) in accordance with following formulas:

$$\dot{u} = \dot{u}_{t+\Delta t/2} = \dot{u}_{t-\Delta t/2} + \ddot{u}_t \cdot \frac{\Delta t_t + \Delta t_{t+\Delta t}}{2} \tag{35}$$

$$u = u_{t+\Delta t} = u_t + \dot{u}_{t+\Delta t/2} \cdot \Delta t_{t+\Delta t} \tag{36}$$

where u_t is a vector of instantaneous velocity and $u_{t-\Delta t}$ and $u_{t+\Delta t}$ are vectors of previous or subsequent displacements.

The software has sophisticated algorithms for complex nonlinear contacts [21], where ongoing simulation model is divided into a selected sequence of m-intervals (where $m \leq t$ and $m_{\min} = 1$). For each time step, the displacement vector u_t is calculated. This vector describes that in following time step, the origin geometry $A0$ is changed to current geometry $A_{t+\Delta t}$ in consequence of the change of displacement vector $u_{t+\Delta t}$ in Eq. (37).

$$A_{t+\Delta t} = A0 + u_{t+\Delta t} \tag{37}$$

In further steps, instantaneous Cauchy stress $\sigma_{t+\Delta t}$ can be expressed using constitutive relations according to Eq. (38) that the algorithm processor expresses as strain change of elements $d\varepsilon = \partial u / \partial X_i$ (i =1,.., 3). Subsequently, a new vector of internal forces for individual nodes is calculated. Variable $t + \Delta t$ is overwritten with t, and the calculation proceeds to the next step.

$$\sigma_{t+\Delta t} = f(\sigma_t, d\varepsilon), \tag{38}$$

The resulting simulation time step Δt is described with Eq. (39) and relates to the calculation time, which is proportional to the size of the smallest element $l_{\min}$, the square root of the material density ρ, and inversely proportional to the square root of elastic modulus E.

The advantage of the explicit method is an order of magnitude faster than explicit method, because the implicit method the time step becomes a quadratic function [21].

$$\Delta t \leq \Delta t^{krit} = l_{\min} \cdot \sqrt{\frac{\rho}{E}} \tag{39}$$

where Δt^{krit} expresses the minimal (critical) time step for the simulation.

Subsequently, the processor for viscoelastic structure expresses Cauchy stress σ_{ij} by the tensor of a nominal stress σ_{ij}^{nom}, which is inversely proportional to vectors extension λ_i (Eq. (40)) as described [22].

$$\sigma_{ij} = \frac{\sigma_{ij}^{nom}}{\lambda_i \cdot \lambda_k} \tag{40}$$

where λ_i expresses vectors of extension to the principal directions and λ_k is a permutation index.

4.2. FEM simulation of mechanical properties of selected samples of PU foam

Simulations were performed in FE software for sample of the PU foam sample. Simulations were performed for a complete assessment of the selected material because the experimental methods cannot give an explanation of behavior and change of the shape, especially under dynamic loading. Model simulations were performed in the following steps:

- an assembling of two models dynamically compressed between rigid plates without initial deformation,
- a creation of the appropriate finite element mesh of the model in preprocessor and import of data file into the environment of PAM CRASH,
- a defining of appropriate initial and boundary conditions,
- an assembly of a nonlinear material model of selected samples,
- evaluating and comparing of the results of simulations in postprocessor.
- assembling of two models dynamically compressed between rigid plates without initial deformation,
- FE model consists of three parts: the test specimen and two plates (moveable and rigid). For each part of the model, structured finite element mesh using a special software Altair 11.0 Hypermesh was created. The finite element mesh was then imported through a text file with the extension *.pc* to the software PAM CRASH. All input parameters of the simulation model (material properties, loads, experimental data, contacts etc.) can be entered and verified in the data file. Assembled FE model is shown in **Figure 20**. The environment of the text file is shown in **Figure 21**.

Applied types and sizes of elements that affect the final time step Δt of the model (Eq. (39)) are shown in **Table 3**.

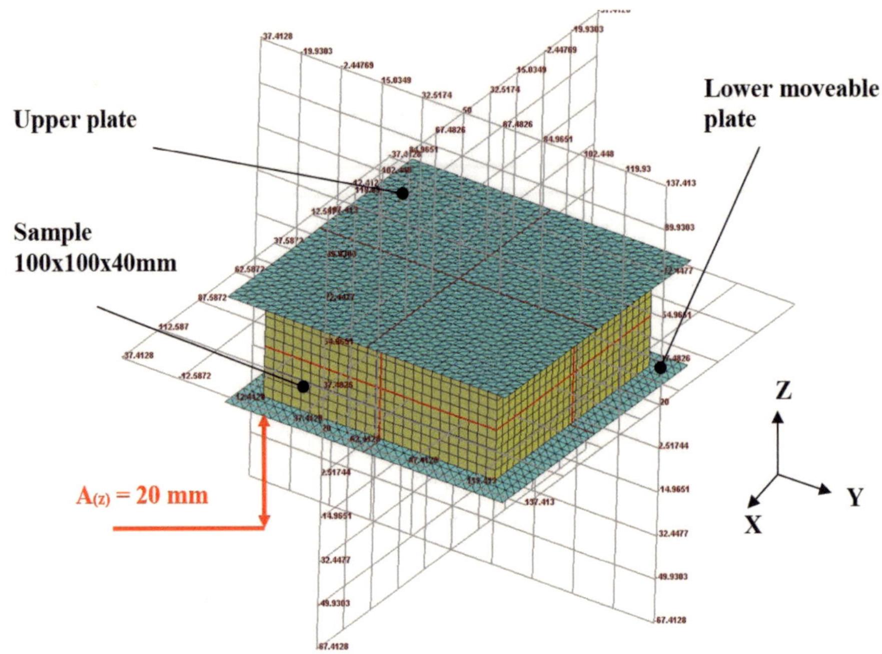

Figure 20. FEM model of sample dynamically compressed without the initial deformation.

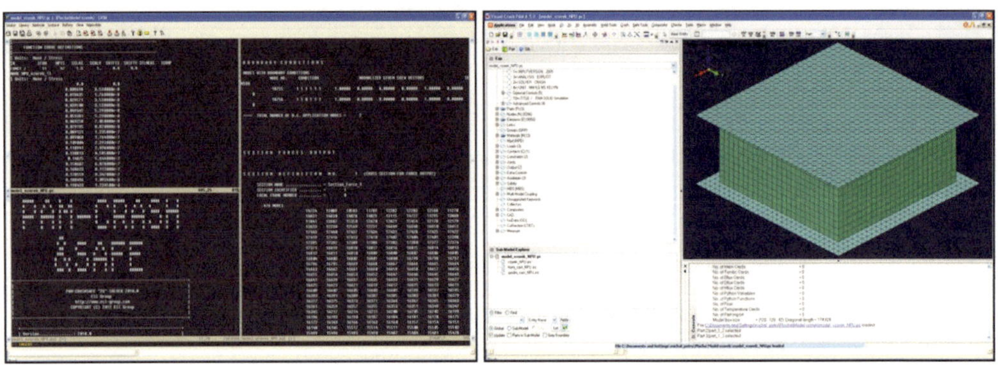

Figure 21. Modification of the input parameters in the FEM model dataset (left); visualization of the simulation model in PAM CRASH (right).

Model	Element type	Element size [mm]	Element number	Friction in contact	Distance between contacts [mm]	Time step Δt [s]
Rigid plates	2D shell	4	3600	0.1	0.5	5.297×10^{-1}
PU foam	3D solid	4	6250	0.1	0.5	0.1471×10^{-4}

Table 3. FEM model dynamically compressed sample.

5. The initial and boundary conditions of the sample compressed with rigid plate

The initial and boundary conditions were defined as in the experiment. They can be summarized in the following points:

- geometric dimensions of the model sample are 100 × 100 × 40 mm,
- bottom plate was defined as perfectly rigid (so-called rigid body) with the translation in the vertical axis Z ($u_z \neq 0$, $u_{x,y} = 0$) with harmonic frequency and amplitude $A(z) = 20$ mm (i.e., 50% deformation) that is defined with Eq. (6),
- upper plate was again defined as rigid body without any translation (fixed) in all directions ($u_{x,y,z} = 0$), and
- contacts were defined between contact surfaces of the sample with plates (**Table 1**).

A solution of contact (Eq. (22)) between two or more parts in the explicit method consists in the fact that the surface of the acting component, so-called *Slave Area* (the plate), pushes on the surface of the second part, so-called *Master Area* (the sample), wherein the contact is calculated between the nodes of two bodies which are in the conjunction [23] (**Figure 22**).

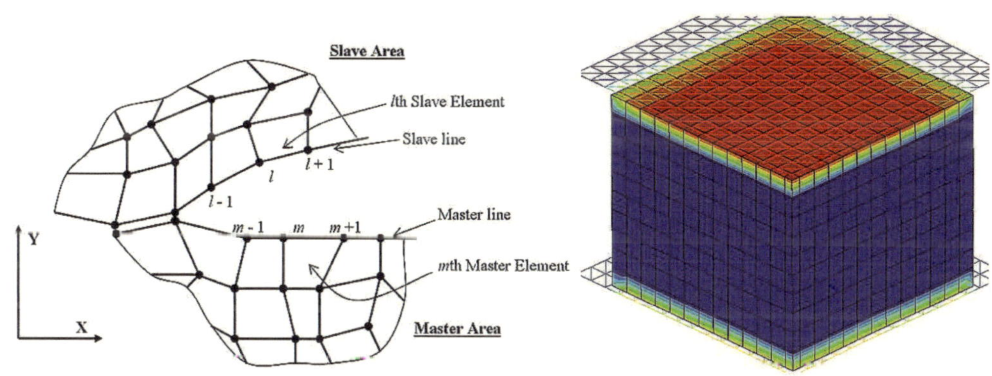

Figure 22. The principle of contact in the model simulations using explicit FEM scheme (left), and resulting model (right).

6. Material model assembling of selected PU foam

Material models in the PAM CRASH chosen to describe the nonlinear behavior of selected samples are prepared from material library, namely:

- Material model describing the mechanical behavior of PU foam has been defined through a nonlinear material model 45—*General Nonlinear Strain Rate Dependent Foam with Optional Energy Absorption*. This material model has already been used and published, for example, in Refs. [3, 9, 12]. Its advantage is to especially allow to assess the influence of rigidity of polyurethane foams depending on the strain rate. The model is based on the rheological behavior of the modified Kelvin model that allows to express Cauchy stress σ_i in the loading axis in accordance with Eq. (41).

$$\sigma_i = E \cdot \varepsilon_i(t) + \eta_t \cdot \dot{\varepsilon}_i(t), \tag{41}$$

where E is modulus of elasticity, $\varepsilon_i(t)$ and $\dot{\varepsilon}_i(t)$express the strain and strain rate in a single direction, and η_t is a damping of the material.

- Another suitable material can be material model 37—*Viscoelastic Ogden Rubber for Solid Elements*—which allows to describe not only viscoelastic but also hyperelastic material properties (suitable for the study of rubber, polymers, fibers, foams, etc.). This is based on the description of the functional dependence of the strain energy density $E(\lambda_1, \lambda_2, \lambda_3)$ defined by Eq. (42), expressing the energy that is required for the structure deformation. Practically, it is analogy to Eqs. (25) and (30).

$$E(\lambda_1, \lambda_2, \lambda_3) = \sum_{i=1}^{3} \frac{\mu_p}{\alpha_p} \cdot \left(\sum_{i=1}^{3} \lambda_i^{\propto_\sigma} - 3 \right) \tag{42}$$

where $i = 1, \ldots, 3,\ E(\lambda_1, \lambda_2, \lambda_3)$is the strain energy density, μ_p and α_p are material constants, and let $\sum_{p=1}^{n} \frac{\mu_p \cdot \alpha_p}{2} = G$, where G is the shear modulus defined by Eq. (43), and $\lambda_i^{\propto_\sigma}$vectors are elongation in principal directions.

$$G = \frac{F}{2 \cdot (1 + v)}, \tag{43}$$

where E is elastic modulus and v is Poisson's ratio.

Using strain energy $E(\lambda_1, \lambda_2, \lambda_3)$, the Cauchy stress in principal directions σ_i can be expressed in Eq. (44).

$$\sigma_i = p_k + \lambda_i^{\propto_\sigma} \cdot \frac{\partial E(\lambda_1, \lambda_2, \lambda_3)}{\partial \lambda_i^{\propto_\sigma}} \tag{44}$$

where p_k is compression stress.

The input material parameters of the simulation model are given in **Table 4**.

Part	Material model	Density [kg m^{-3}]	Initial module E [MPa]	Poisson ratio γ [–]	Damping coefficient
Rigid plates	Linear elastic	7850	210,000	0.3	–
PU foam	mat. 45	50.16	2.6	–	0.2

Table 4. Material properties of FEM model dynamically loaded sample.

7. Results of simulations

The input signal and a material response are shown in **Figure 23**. The results of simulations for frequency 5 Hz are shown in **Figure 24**. In comparison with the real experiment, simulated values have a high correlation coefficient (0.961) up to 37.5% deformation (compression of 15 mm). The correlation coefficient between the model and the real measurement when compressed to 50% deformation exhibits a correlation of 0.932. The results of dynamically compressed samples against the rigid plate without the initial deformation with parameters according to **Table 4** are in good agreement with experiments.

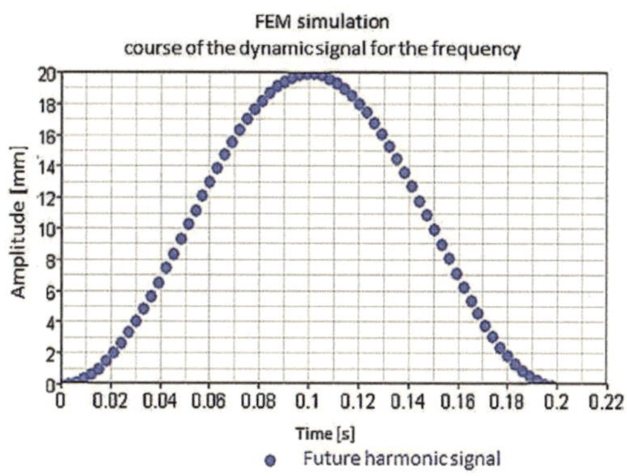

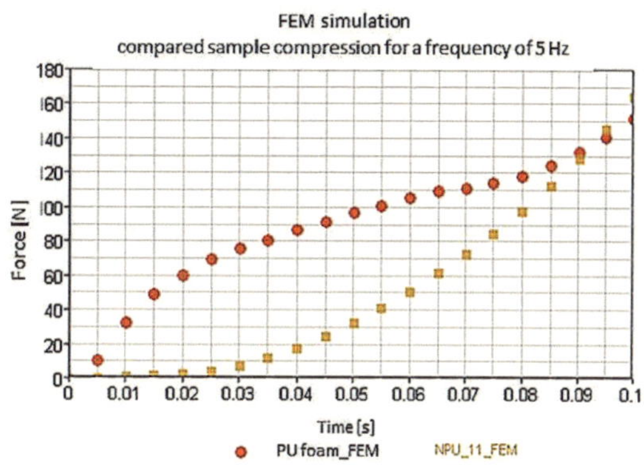

Figure 23. FEM model: excitation signal (left), and the response of the material on compression (right).

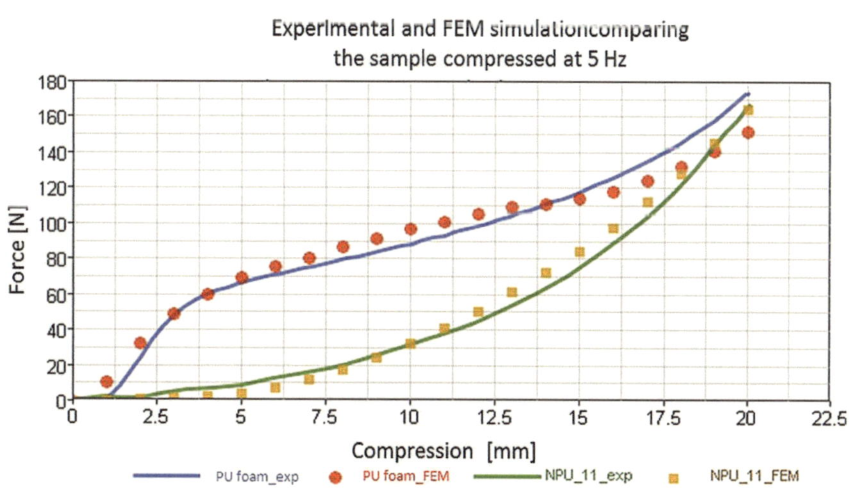

Figure 24. Comparison of the courses of experiment and FEM models.

The distribution of strain in the *X* and *Y* directions (plane perpendicular to the axis of compression) at a 37.5% deformation of the sample showed (**Figure 25**) that a sample of the PU foam in these directions is significantly deformed, because in these areas, there is highest stress. The maximum value of displacement vectors is 6404 mm, which is approximately 15% deformation of the sample. This leads to the fact that the structure is substantially pushed out from the sample. In contrast, non-polyurethane sample no. 11 is not practically in the plane perpendicular to the axis of compression deformed (**Figures 26–28**), because the maximum value of the displacement vectors is 0.034 mm.

Strain distribution also influences the value of the maximum principal stress during compression, as shown in **Figures 26–28**.

Results of the main stress shown in **Figure 28** indicate that material of the PU foam sample is pushed out already at 12.5% deformation. This phenomenon confirms that the PU foam with increasing strain rate increases the stiffness and the foam is pushed out. These results were further studied and tested in Refs. [3, 12]. The simulation of contact pressures has shown that the strain of the foam in directions perpendicular to the direction of compression leads to uneven stress distribution, which is reflected by the uneven distribution of contact pressures; however, these results cannot be obtained experimentally under dynamic loading. That is why

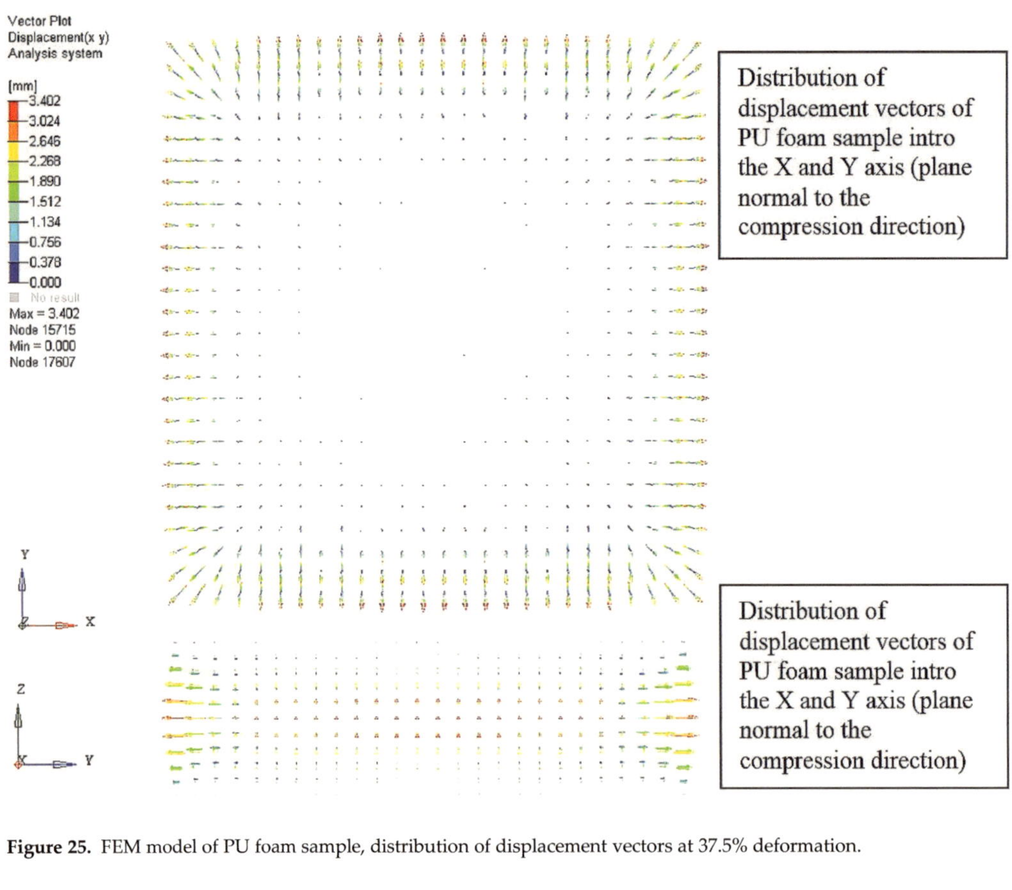

Figure 25. FEM model of PU foam sample, distribution of displacement vectors at 37.5% deformation.

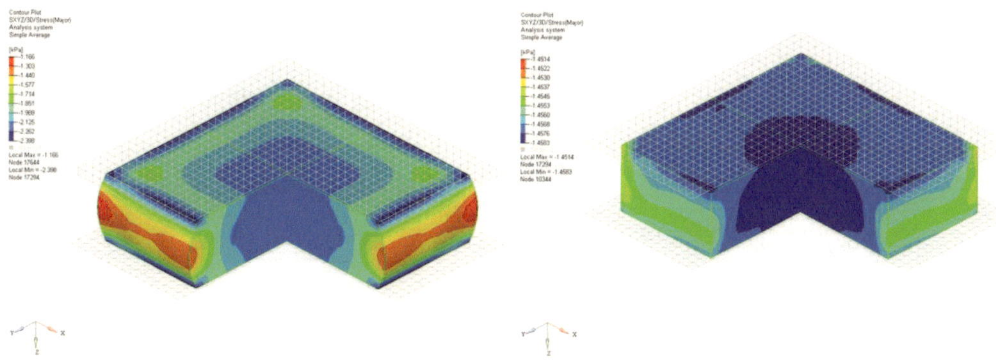

Figure 26. Deformation 37.5%: FEM model of PU foam sample.

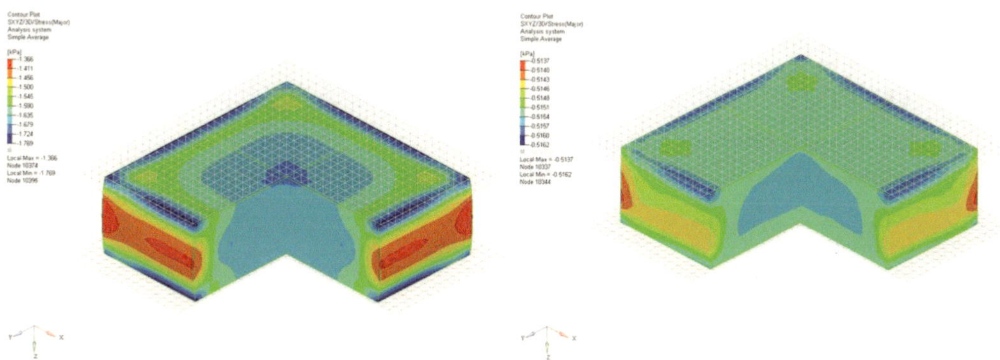

Figure 27. Deformation 25%: FEM model of PU foam sample.

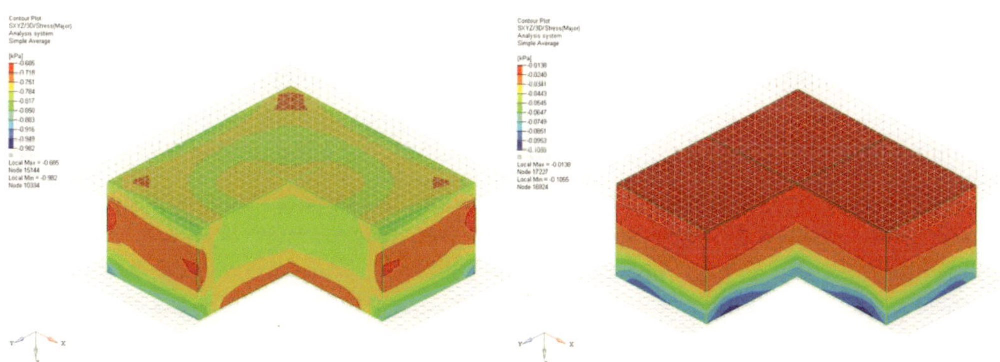

Figure 28. Deformation 12.5%: FEM model of PU foam sample.

the modeling is a suitable tool for obtaining of information that is not possible to achieve by the experiments (**Figure 29**).

The results of loading at frequency 5 Hz and 37.5% deformation are summarized in **Table 5**.

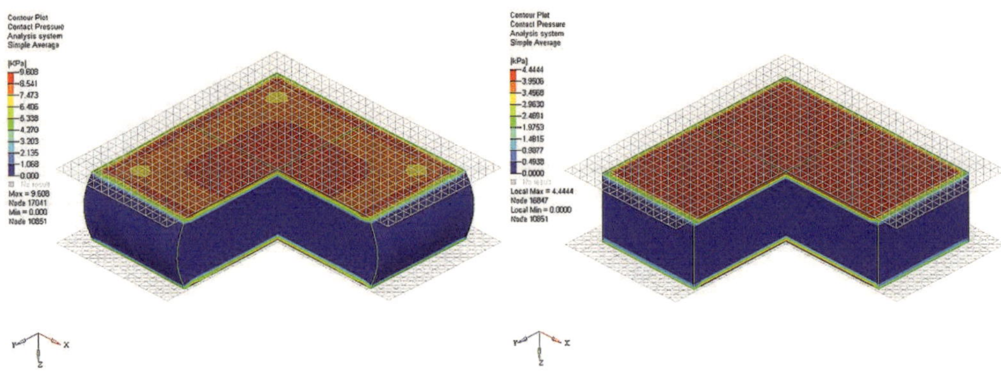

Figure 29. FEM model: comparison of contact pressures distribution at 37.5% deformation.

Sample	Deformation [%]	Principal stress [kPa]	Total stress [kPa]	Contact pressure [kPa]
PU foam	37.5	2.398	2.422	9.608

Table 5. Results of stress in dynamically compressed sample.

8. Conclusion

For static and dynamic measurements of samples, experimental devices were designed and implemented. Results showed that mechanical properties of PU foams are dependent on the strain rate, because the air in the initial stage cannot escape from the structure under dynamic loading, thus the air participates on the resulting foam properties. These results in increase of the main stress in the peripheral parts of the PU foam sample, as shown in the simulation of the sample loaded with harmonic signal. From the results of measurement of the relaxation, it was observed. Using the FEM simulation, the distribution of contact pressures was also compared. It was showed that the strain of the PU foam in directions perpendicular to the direction of compression leads to uneven stress distribution, which is reflected by the uneven distribution of contact pressures.

The described behavior influences properties of products containing PU foams. This behavior is significantly reflected in the construction of car seats. For example, headrests are used to capture the head in the rear impact (so-called whiplash). The energy absorption may be insufficient due to the setting of the headrest. If the backrest is fully inserted, a significant deflection of the spine in the area of the cervical vertebrae occurs, while if the headrest is correctly set, the spine is not deflected. Also, a dependence of material stiffness on strain rate has certain influence. The solution could be not only in the modification of the geometry and the anatomical design of the seat, but also in the change of the headrest stuff material which is not so dependent on the strain rate. The solution could be in a usage of foams with a highly porous structure and foams in the combination with fibrous structures that do not exhibit strong dependence of stiffness on the strain rate.

Furthermore, it has been shown that the PU foam is not capable to transmit tensile stresses. If the foam is under tensile stress, the stiffness will be very high with very low ability to deform. This means that low deformation and also the force cause a tearing of the cell structure. If the torn foam is further stressed, complete destruction occurs. This is especially important for mounting and assembling, such as inserting a seat foam, backrest, or mattress into the cover. This problem is further accentuated by relatively high surface friction of the foam to other materials. Therefore, it is suitable to use an interlayer that reduces the friction between the surfaces, e.g., interposed polyethylene film, which is removed after assembly or layer, which is inserted permanently. On the other hand, the foam that is already torn can transmit a pressure load without significant impact if the foam is placed in a cover or box that does not allow its distortion in a direction that is perpendicular to acting load.

Also, it has shown that molded polyurethane foam has a qualitatively different surface compared to the inner structure of the foam. The foam surface is closed, its porosity is very low, and its character approaches the integral foam. The surface also exhibits other mechanical properties, especially significantly higher stiffness in all kinds of loadings. This affects the load behavior, which is further influenced by the low permeability of the foam surface. As mentioned, the response to strain rate is a specific characteristic of foams and it is related to the cellular structure, especially its openness. If the surface is closed or only partially permeable to air, it influences the stiffness of the foam at a fast compression. The air is not able to immediately escape from the structure, and due to limited air compressibility, the rigidity of the foam increases. This can be seen especially in the first loading cycle, which is a crucial for impact during the crash when the human body is pushed into the foam. Therefore, it is appropriate to finish the surface in a suitable way, for example, by a grinding, perforating, grooving, etc. However, it should be noted that PU foam is invaluable material for static or quasi-static type of loadings. In particular, it is necessary to highlight behavior under compression loading. The strain-stress curve course is characterized by almost flat shape, so the force increases very slowly in large range of the deformation. This leads to the fact that the foam provides high comfort, because the foam very easily surrounds the body and reduces the contact pressure. This is not practically achievable for other materials and therefore PU foam is a material that is not easy to replace. Above all, PU foams exhibit balanced mechanical, physical, and thermal properties at low bulk densities and low weight. Also prduction costs are very low. That are the reasons for their use.

Author details

Michal Petrů[1]* and Ondřej Novák[2]

*Address all correspondence to: michal.petru@tul.cz

1 Institute for Nanomaterials, Advanced Technology and Innovation, Technical University of Liberec, Liberec, Czech Republic

2 Department of Nonwovens and Nanofibrous Material, Textile Faculty, Technical University of Liberec, Liberec, Czech Republic

References

[1] Mills NJ. Polymer Foams Handbook: Engineering and Biomechanics Applications and Design. Butterworth-Heinemann, UK; 2007

[2] Woldesenbet Em, Peter S. Volume fraction effect on high strain rate properties of syntactic foam composite. Journal of Material Science. 2009;**44**(6):1528–1539

[3] Petrů M, Novák O, Prášil L. Reduce of head injuries during whiplash by the help of materials with independent strain rate. Journal of Rehabilitation Medicine. 2011;**43** (Suppl. 50):28–29

[4] Cirkl D. Measurement of mechanical properties of polyurethane foam in vacuum. In: Academy of Science of the Czech Republic, editor. Dynamics of Machine; 6–7 February 2007; Prague. Prague: Academy of science of the Czech Republic; 2007

[5] Landrock AH. Handbook of Plastic Foams. USA: Library of Congress New Jersey; 1995

[6] Foye RL. Compression Strength of Unidirectional Composites. Columbus, USA: American Institute of Aeronautics and Astronautics, Structural Composites Group; 1996

[7] Bareš RA. Kompozitní materiály. SNTL Praha; 1988

[8] Mills NJ. The high strain mechanical response of wet Kelvin foams open cell foams. International Journal Solids Structure. 2007;**44**:51–65

[9] Petrů M, Petřík J. Systems to optimize comfort and developments of car seat. Acta Technica Corviniensis – Bulletin of Engineering, Annals of Faculty Engineering Hunedoara. 2009;**4**: 55–59

[10] Rush KC. Energy-absorbing characteristics of foamed polymers. Journal of Applied Polymer Science. 1970;**14**:1133–1147

[11] Neilsen MK, Morgan HS, Krieg RD. A Phenomenological Constitutive Model for Low Density Polyurethane Foams. SANDIA Report; 1987

[12] Petrů M, Novák O. Analysis and testing of mechanical properties of polyurethane foam and materials for nonpolyurethane car seat cushions. In: Praha CZU, editor. 4th International Mechanical Engineering Forum Prague; 2011; Prague. Prague: CZU Praha; 2011. pp. 135–149

[13] Rush KC. Load compression behavior of flexible foams. Journal of Applied Polymer Science. 1969;**13**:2297–2311

[14] Schwaber MD, Meinecke AE. Energy absorption in polymeric foams. II. Prediction of impact behavior from instron data for foams with rate-dependent modulus. Journal of Applied Polymer Science. 1971;**15**(10):2381–2393

[15] Nagy A, Ko WL, Lindholm US. Mechanical behavior of foamed materials under dynamic compression. Journal of Cellular Plastics. 1974;**10**(3):127–134

[16] Sinha SC, Mitchell JO, Lim GG, Chou CC. Constitutive modelling of energy absorbing foams. 1997; SAE paper no. 940880

[17] Zhang J, Kikuchi N, Li V, Yee A, Nusholtz G. Constitutive modeling of polymeric foam material subjected to dynamic crash loading. International Journal of Impact Engineering. 1988;**21**(5):369–386

[18] Liu Z, Zhang E, Ji Z. Simulation and experimental study of human Riding comfort in dynamic man-automobile system. Lecture Notes in Computer Science. 2008;**5314**(2008): 577–587

[19] Okrouhlík M, Höschl C, Plešek J, Pták S, Nadrchal J. Mechanika poddajných těles, numerická matematika a superpočítače. Praha: ÚT; 1997

[20] Kleiven S. A parametric study of energy absorbing foams for head injury prevention. In: Proceedings of the 20th International Technical Conference on the Enhanced Safety of Vehicles (ESV); 2007; Lyons France; 2007

[21] Belytschko T, Liu WK, Moran B. Nonlinear Finite Elements for Continua and Structures. Chichester, UK: John Wiley & Sons, Ltd; 2000

[22] Flanagan DP, Taylor LM. An accurate numerical algorithm for stress integration with finite rotations. Computer Methods in Applied Mechanics and Engineering. 1987;**62**:305–320

[23] Clinckemaillie J, Galbas HG, Kolp O, Thole CA, Vlachoutsis S. High scalability of parallel PAM-CRASH with a new contact search algorithm. Lecture Notes in Computer Science. 2000;**1823**:439–444

Polyol Containing Boron Atoms as a Compound which Reduces Flammability of Rigid Polyurethane-Polyisocyanurate Foams

Joanna Paciorek-Sadowska

Additional information is available at the end of the chapter

http://dx.doi.org/10.5772/intechopen.69585

Abstract

The article presents results from multidisciplinary research conducted with the purpose of obtaining new material which reduces flammability and is used for the production of rigid polyurethane-polyisocyanurate foams (PUR-PIR). During the preparation of the foam recipe, special attention was paid to the ability to lower the flammability and the improvement of durability and thermo-insulation properties of rigid foams by using the new, non-halogen polyol. The new boron compound helped obtaining fire-safe PUR-PIR foams, which are characterized by the reduced amount of total exhausted heat, longer time to full combustion, lower amount of produced carbon dioxide and oxide, increased oxygen index and lower maximum burning temperature in shorter time than the reference foam.

Keywords: polyol, flammability of foams, rigid polyurethane-polyisocyanurate foams, properties of the rigid foams

1. Introduction

Polyurethanes are polymer materials that have been significantly developed, especially in the last 10 years worldwide. Many facilities produce new raw materials and new polyurethane materials, keeping in mind the principles of sustainable development. Polyurethane advancement started over 75 years ago with the first patent claim from Otto Bayer. However, the first reaction resulting in polyurethane bond was conducted by Wurtz, using diethyl sulphate and potassium isocyanate [1, 2]. Currently, polyurethanes are one of the largest growing polymer groups. The interest in them is related to their specific characteristics which can be widely modified. By changing the raw materials, their correlating volume ratio and by selecting appropriate processing conditions, it is possible to obtain solid, porous, composite, leather-like and biodegradable materials, and elastomers, glues, fibres, adhesives and many more. Rigid polyurethane foams have a special place among them. They have low apparent density and excellent mechanical properties. That is why they are used in many fields, for example, construction, automotive, furniture, shoe and packaging industries [3–6]. The superb thermo-insulating properties of the rigid foams are the main reason for their wide usage. They are the best thermo-insulating material used in construction and refrigeration. The main drawback of the foams, and one that may limit the expansion of polyurethanes for new applications, is their flammability. It is an important issue, especially when using polyurethanes on large surfaces and in public buildings [7–20]. Lowering their flammability requires multidisciplinary solutions during the stage of designing the chemical structure and antipirenes. Usually, phosphorous, nitrogen, boron and halogens compounds are introduced to polyurethane materials. Nowadays, however, there is a worldwide tendency to withdraw the antipirenes produced based on chlorine and bromine because of the high toxicity during their thermal decomposition. The addition of large amounts of antipirenes in order to obtain the desired non-flammable effect causes technological issues and significantly decreases the physicomechanical properties and dimensional stability of the produced materials [21–24].

Based on the conducted research, the author stated that in order to omit the inconvenience related to the use of halogen antipirenes, the foamed polyurethanes with high-flame resistance can be produced by using the new polyurethane composition which contains the new boron-nitrogen polyol which decreased flammability.

2. Experimental part

2.1. Characteristics of raw materials

The polyether with trade name Rokopol RF-55 (product of oxypropylation of sorbitol LOH = 495 mg KOH/g, produced by NZPO 'Rokita', Brzeg Dolny, Poland) and Ongromat 20–30 (technical polyisocyanate whose main component is diphenylmethane 4,4′-diisocyanate, made in Hungary) were used to prepare the rigid PUR-PIR foams. The density of Ongromat 20–30 at a temperature of 25°C was 1.23 g/cm^3, its viscosity was approx. 200 mPa s. The polyisocyanate contained 31.0% of isocyanate groups. It was characterized according to the ASTM D 1638-70 standard.

An anhydrous potassium acetate in the form of 33% solution in diethylene glycol (Catalyst-12, POCh Gliwice, Poland) and amine catalyst in the form of 33% solution of triethyleneamine in dipropylene glycol (DABCO, Hondrt Hülls, Germany) were applied in the foam composition.

The polysiloxanepolyoxyalyleneoxydimethylene copolymer characterized by the boiling point of 150°C at 1013 hPa and ignition temperature of 90°C (Niax Silicone L 6900, Witco Corp., USA) was used as a surfactant.

The porophor was carbon dioxide formed in situ in reaction of isocyanate with distilled water.

Moreover, tri(2-chloro-1-methyl-ethyl) phosphate (Antiblaze, Albright and Wilson, UK) was introduced into foams.

Boran tri[N,N′-di(metylenooksy-2-hydroksyetylo)mocznika] obtained from the Faculty of Chemistry, Technology of Polymers and Ecotechnology of Bydgoszcz Academy was applied as a modifier for the preparation of foams.

2.2. Synthesis of boronitrogen polyol from boric acid and di(hydroxymethyl)urea

The reaction for obtaining boronitrogen polyol was conducted in an oxygen atmosphere, at the temperature of 131°C, using glass equipment: three-neck flask placed in electric heating mantle. The flask was equipped with Deana-Stark head, thermometer and mechanical stirrer. The flask contained 61.81 g (1 mole) of boric acid (H_3BO_3 as a white powder with a molecular mass of 61.81 g/mol, produced by POCh in Gliwice), 212 g (2 moles) of xylene (a mixture of o- and p-isomers, produced by Chempur in Piekary Śląskie, with water content above 0.1% and density of around 0.860–0.866 g/cm^3), 876 g (3 moles) of N,N′-di(methylenoxy-5-hydroxy-pentylo)urea (yellow-brown liquid, with 1193 g/cm^3 density, produced in the Chemistry and Technology of Polyurethane Department in Bydgoszcz). The flask content was kept at a boiling temperature for 510 min with constant and intense stirring. When the reaction ended and the stirring stopped, the reactive mixture was separated into an upper clear layer (xylene) and a light-blue bottom layer (new polyol). After cooling, the separated bottom layer was kept for 1 h in the temperature of 130°C in vacuum dryer under the pressure of 0.320 kPa to remove any solvent and water residue.

As a result of the conducted reaction, borane tri[N,N′-di(methylenoxyethylentio-2-hydroxyethyl) urea] was produced (**Scheme 1**).

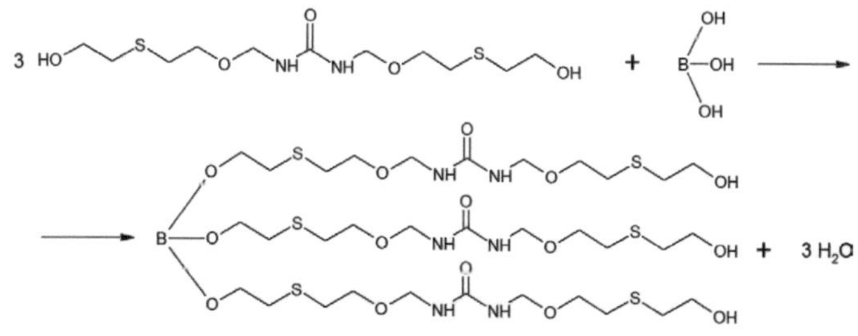

Scheme 1. Reaction for obtaining borane tri[N,N′-di(methylenoxyethylentio-2-hydroxyethyl)urea].

Examining the physicomechanical properties of the new compound was the next step. The results of the examination helped determining the usability of the obtained borane as a polyol for the production of rigid polyurethane-polyisocyanurate foams. A polyol is a basic component of the polyurethane composition and it determines the properties of produced materials in a significant way. **Table 1** describes the physical properties of the new polyol.

The viscosity of the obtained borane is 283 mPa s; however, the density is 1300 kg/m^3. The viscosity value of the new polyol does not exceed the viscosity value of industrial polyols (15,000 mPa s). Higher viscosity values of the raw materials used in polyurethane production

Polyol's name	Properties						
	Hydroxyl number (mgKOH/g)	Density (20°C) (kg/m³)	Viscosity (20°C) (Pa s)	Mole mass (g/mole)	Colour	Nitrogen volume (%mass.)	Boron volume (%mass.)
Borane tri[N,N'-di(methylenoxyethylentio-2-hydroxyethyl)urea]	430	1300	283	546	Colourless	9.7	1.3

Table 1. Borane properties.

are not advisable due to the technological limitations of standard equipment used for PUR processing, available on the market.

The structure of the new polyol indicated the presence of boron and nitrogen that have a significant influence on flammability. Based on the conducted calorimeter analysis with carminic acid, it has been noted that the boron value is 1.3% of the mass, and nitrogen value is 9.7% of the mass.

Hydroxyl number is an elementary parameter of a polyol which is important during the calculation of polyurethane composition. That is why it has been determined for the obtained boroorganic compound. The hydroxyl number is 380 mgKOH/g and is similar to its calculated theoretical value. It also proved that the compound in an interesting material for the production of rigid PUR-PIR foams.

The presence of water is also an important characteristic of polyols used for obtaining polyurethane foams. The amount of water needs to be precisely determined in those compounds because it can have a substantial influence on the foaming process during the reaction with the polyisocyanate. The amount of water in the obtained borane is lower than 0.1%. This value does not interfere with the process of foam production and does not have to be included in the recipe. The structure of the obtained borane compound was determined based on proton nuclear magnetic resonance spectrum ^{1}HNMR, using chloroform as the dissolvent. The results of ^{1}H NMR analysis of the polyol are represented in **Table 2** and **Figure 1**.

The presence of characteristic groups for obtained compounds was confirmed with the analysis of the spectrum of new boron compounds, using spectroscopy in infrared. The IR spectrum shows bands with frequencies typical for groups present in the new polyol. The results are presented in **Figure 2**.

Hydrogen atom position	Chemical shift (ppm)]	Number of hydrogen atoms
1	3689–3609	3
2, 6, 11, 15	4693–4308	24
3, 4, 5, 12, 13, 14	1426–1294	36
8, 9	6995–6532	6

Table 2. Assigning signals in ^{1}H NMR spectrum to particular protons for borane tri[N,N'-di(methylenoxyethylentio-2-hydroxyethyl)urea].

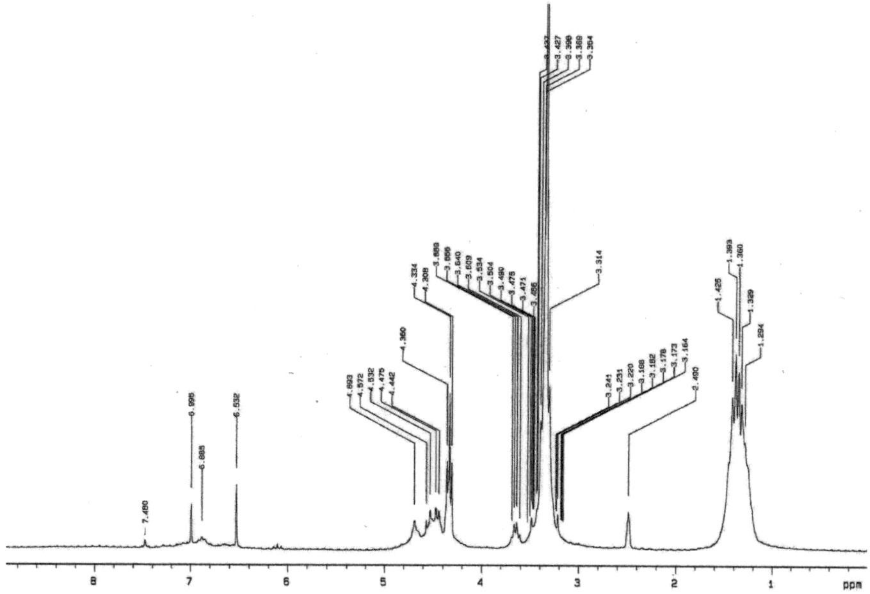

Figure 1. ^{1}H NMR spectrum of borane tri[N,N′-di(methylenoxyethylentio-2-hydroxyethyl)urea].

The IR spectrum shows, among others, characteristic bonds representing frequencies of NH-CO-NH bonds, at 1648–1654 cm^{-1}, and in the 1290–1336 cm^{-1} for B-O bonds. Also, a band typical for hydroxyl group –OH appeared in the new polyol at the 2949–3556 cm^{-1} range which are also confirmed by the results of the determined hydroxyl groups.

2.3. Synthesis of rigid PUR-PIR foams in 18-dm^3 mould

Foamed polyurethanes are usually produced based on a so-called polyurethane system which comprises two or more components that react during mixing and give the desired product. This article focuses on the examination of polyurethane-polyisocyanurate foams produced using two-component A+B system. The A component was obtained by mixing polyols (Rokopol RF-55 and N,N′-di(hydroxymethyl)urea derivative), catalysts, surface-active agent, and porophor. Component B on the other hand was polyisocyanate. Foams produced with this system are very simple and the process is not too energy intensive. However, the development of processing technology of foamed polyurethanes encountered a serious issue regarding the requirements for environmental protection. Developing a foaming agent that would be environmentally safe and, at the same time, would help producing materials with valuable properties is a great challenge for polyurethane manufacturers. In the presented research, carbon dioxide was used as the foaming agent which was exhausted during the reaction or water with isocyanate groups. Despite a common opinion that carbon dioxide slightly deteriorates the application properties of a polyurethane foam (heat conductivity, durability and brittleness), initial research on using the new polyol in the polyurethane composition shows a promising final product with beneficial properties. The basis for calculating the polyurethane recipe was the hydroxyl number

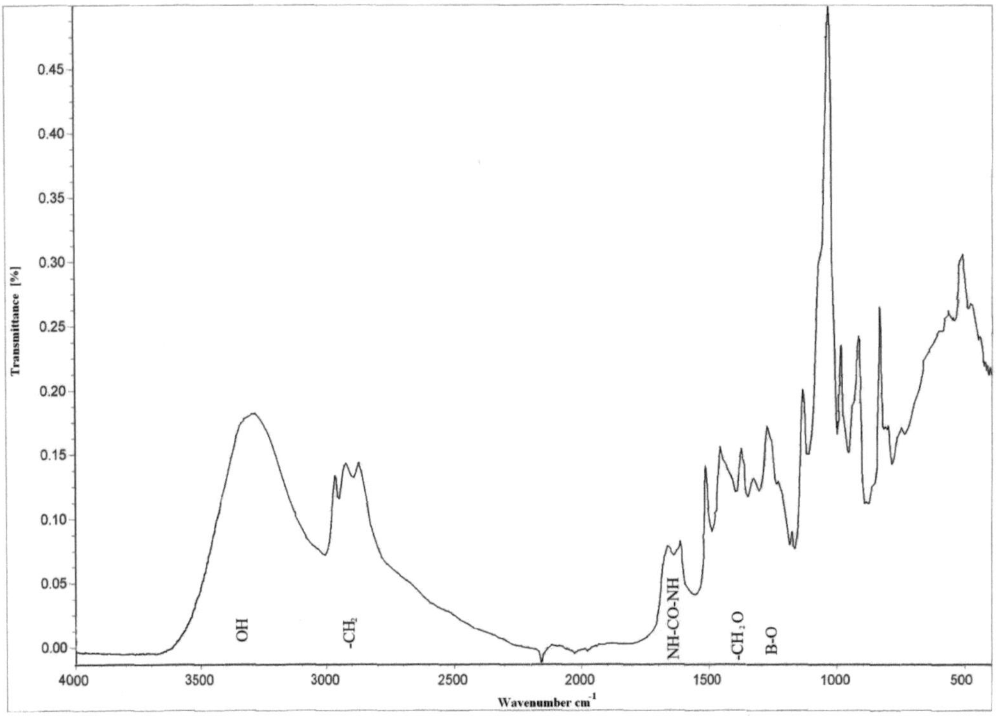

Figure 2. IR spectrum for boroorganic compound recorded by KBr technique.

of the used polyols. Then, the values of auxiliary compounds which usually do not contain hydroxyl groups were determined, for example, catalysts, non-reactive flame retardants and surface-active agents. They were described in weighted portions in relation to 100 weighted portions of the polyol. The amount of isocyanate was selected with regard to the ratio of isocyanate groups to hydroxyl groups, which was set to 3:1 for rigid polyurethane-polyisocyanurate foams. The calculated amount of isocyanate was increased by the isocyanate mass needed to conduct a reaction with water, as a result of which CO_2 was exhausted—the gas which foams the reactive mixture. The recipes for obtained foams are described in **Table 3**.

When starting with the rigid PUR-PIR foam production based on the determined recipes (**Table 3**), polyisocyanate was measured in the first 1-dcm^3 polypropylene vessel, and polyol component polyol with added auxiliary substances was measured in the second vessel. The polyol was thoroughly mixed with other components using electric stirrer with 1800-rpm rotation speed.

The polyisocyanate was combined with the polyol component, stirred for 10 s and both were poured into a mould where the foam's rising process was observed. An open mould was used for the examination which enabled a so-called free-rising process.

During the first stage, the reference K0 foam was obtained, which did not contain any amount of the polyol, and then foams marked from K1 to K5 were produced by adding from 0.1 to 0.5 R of the polyol, respectively.

	Unit	K0	K1	K2	K3	K4	K5
Rokopol RF-55	R	1	0.9	0.8	0.7	0.6	0.5
	g	55.65	50.98	44.78	31.35	18.81	9.40
Borane tri[N,N′-di(methylenoxyethylentio-2-hydroxyethyl)urea]	R	0	0.1	0.2	0.3	0.4	0.5
	g	0	6.26	12.52	18.78	25.04	31.31
Silicone L6900	g	4.7	4.7	4.7	4.7	4.7	4.7
DMCHA	g	2.8	2.8	2.8	2.8	2.8	2.8
Catalyst 12	g	6.5	6.5	6.5	6.5	6.5	6.5
AntiblazeTMCP	g	46.9	46.9	46.9	46.9	46.9	46.9
Water	R	0.7	0.7	0.7	0.7	0.7	0.7
	g	3.15	3.15	3.15	3.15	3.15	3.15
Ongromat 30-20	R	3.0	3.0	3.0	3.0	3.0	3.0
	g	250.6	250.6	250.6	250.6	250.6	250.6

Table 3. The recipes for rigid PUR-PIR foams.

During the polyurethane-polyisocyanurate foam synthesis, the foaming process of the reactive mixture was monitored, and with the help of a stopwatch, the start time, expansion time and gelation time were measured.

Start time is the time measured with a stopwatch from the moment of mixing the components, up to when the foam reached a 'cream state'. It is indicated by the increasing volume of the foam.

Rising time is the time measured with a stopwatch from the moment of mixing the components, up to reaching the maximum volume of the foam.

Gelation time is the time measured with a stopwatch from the moment of mixing the components, up to the moment when the free surface of the foam stops sticking to a clean glass rod.

2.4. Methodology of rigid PUR-PIR foam production

The examination of apparent density was conducted after 24 h of seasoning of the samples in room temperature, according to ISO 845-1988 standard. The apparent density was determined for all the examined samples. The samples were measured with an accuracy up to 0.01 mm and weighted with an accuracy up to 0.01 g.

The brittleness of examined foams was determined according to the ASTM C-421-61 standard. Based on that standard, the brittleness was measured as a percentage mass loss of 12 foam cubes (square cubes with 25 mm sides), during an examination in normalized apparatus, in relation to the initial mass. The apparatus for measuring PUR foam brittleness was a cubical box made out of oak wood, with dimensions of 190 × 197 × 197 mm, rotating along its axis with the speed of 60 rounds per min. The box was filled with 24 oak cubes with the dimensions of 20 × 20 × 20 mm.

The determination of water absorption was conducted in accordance with DIN 53433 standard. This method measures the hydrostatic buoyancy of the sample with the dimensions of 150 × 150 × 25 mm, submerged in distilled water for 24 h.

Thermo-insulating properties of the produced foams were determined by measuring the heat conductivity coefficient λ. Samples with dimensions of 200 × 200 × 25 mm were used for the test. FOX 200 apparatus by Lasercomp was used during this examination. It enables the determination of λ-value in the range of 20–100 mW/(m K). The measurement chamber needs to be fully filled to conduct proper examination.

The heat conductivity coefficient measurement method measures the amount of heat transported through the sample material in a unit of time during the determined heat flow, that is, when the temperature difference is measured on the opposite sides of the examined sample. The measurements are performed in series, in 0.5-s intervals.

Compressive strength was determined by the use of general-purpose strength machine (Instron 5544). The peel and flanks of foam were cut off and the cubic samples were cut out (side of 50 ± 1 mm). Then, the samples were subjected to compressive strain by 10% according to the direction of foam expansion.

The content of closed cells was determined in compliance with PN-ISO 4590 standard, using defect-free samples with the dimensions of 100 × 25 × 25 mm. The method utilizes the Boyle–Mariotte law. It determines a relative pressure decrease based on calibrated volume patterns, measuring the difference on the scale of a manometer with one arm opened to the atmosphere.

The retention (remains after burning) of the PUR-PIR foams was examined according to ASTM D3014-73 standard, by performing the vertical test. The apparatus used for the burning examination using the vertical test had a vertical chimney with 300 × 57 × 54-mm dimensions. Three of the walls were made out of tin and one was a removable glass wall. The test was conducted on six samples with dimensions of 150 × 19 × 19 mm. Before burning, the samples were weighted with an accuracy up to 0.0001 mm and placed inside the chimney. The glass was placed in place and flame was introduced to the samples from propane-butane burner for 10 s. Then, the burner was moved away and the times of free burning and retention were measured with a stopwatch in the vertical test. Retention was calculated using Eq. (1):

$$R_e = \frac{m}{m_0} \times 100\% \tag{1}$$

where R_e is the retention; m_0 is the sample mass before burning (g), m is the sample mass after burning (g).

Using the cone calorimeter, the examination of flame and smoke parameters was conducted for the produced PUR-PIR foams based on the methodology described in the ISO 5660-1:2001 norm.

Normalized samples with dimensions of 100 × 100 mm were subjected to heat radiation. During the examination, the following parameters were recorded: the time needed to initiate the burning process, thermokinetic parameters, that is, heat exhaustion rate and the total amount of exhausted heat, and selected toxic and smokegenic properties. The thermokinetic values were measured using the theory of oxygen usage calorimetry, which states that out of 1 g of used oxygen, there are around 13.1 kJ of heat produced with accuracy up to ±5%. The examination was conducted for material samples placed horizontally and the burning reaction was initiated by combustion. The material samples were exposed to heat radiation of 30 kW/m^2. The examination ended after the burning process faded completely. The temperature distribution during burning was measured using a Vigo V-20E2-25 thermal imaging camera equipped with HgCdTe thermoelectrically cooled detector. The measurement conditions were compliant with conditions for determining the oxygen index (OI).

The oxygen index (OI) was determined in compliance with the methodology described in the ASTM D 2863-1970 standard. It measures the boundary concentration of oxygen in oxygen and nitrogen mixture, sufficient for sustaining burning of a sample with the dimensions of 150 × 13 × 13. The oxygen index was calculated in percentile value according to Eq. (2)

$$OI = \frac{O_2}{O_2 + N_2} \times 100\% \tag{2}$$

where O_2 is the volumetric flow rate of oxygen for the boundary volume (m^3/h), N_2 is the volumetric flow rate of nitrogen for the boundary oxygen volume (m^3/h).

The examination of cell structure was performed using HITACHI S-4700 scanning electron microscope (SEM) with NORAN Vantage microanalysis system. The reference sample (containing petrochemical polyol) and the modified sample with the largest amount of boron polyol were tested.

IR spectroscopy was used to identify characteristic groups present in PUR-PIR. The polyurethane foams were milled in Janetzky's mill before they were tested with the IR method. The analysis of the milled samples was performed with KBr technique on Brucker spectrophotometer in the range from 200 to 4000 cm^{-1}.

3. Examination results and their analysis

3.1. The influence of raw materials on the density and processing times of the obtained foams

The process of creating PUR-PIR foams is very complicated from the chemical point of view. Additionally, there may be technological issues which make the examination of the foaming process of the PUR composition necessary to obtain products with beneficial properties. The foaming process is an important stage in the production of polyurethanes. Here, the cell structure is formed which significantly influences the application properties of the material. During rigid PUR-PIR foams production, processing times (start time, rising time and gelation time) have been determined. Based on the conducted research, different times can be observed of the foam surface losing its adhesiveness. It is dependent on the type of medium (air, mould wall) that the rising foam's surface is touching. The air causes the foam to lose its

Foam symbol	Start time (s)	Rising time (s)	Gelation time (s)	Density (kg/m³)
K0	8	15	31	33.4
K1	6	14	27	40.3
K2	5	10	25	40.6
K3	4	10	21	40.8
K4	4	9	20	41.3
K5	3	8	16	41.5

Table 4. Processing parameters and apparent density of rigid PUR-PIR foams.

adhesiveness the quickest. Apparent density was also measured for the rigid foams and the results of all tests are represented in **Table 4**.

The processing times of modified foams were reduced by about 50% in comparison to the reference foam. The start time for P0 reference foam was 8 s; however, for the foam containing 0.5 R of the boron polyol, it was 3 s. Similar changes were observed for rising and gelation times. This shows a higher reactiveness of the boron-nitrogen polyol in comparison to the industrial one.

Apparent density is one of the most important factors that determine the mechanical properties of rigid PUR foams. From the economical point of view, it is beneficial to produce materials with the lowest possible apparent density. Nevertheless, the apparent density of a PUR foam is in close correlation with its thermo-insulation properties, mechanical properties and dimensional stability. That is why the most commonly used rigid polyurethanes have the density values in the range of 30–60 kg/m^3. By using the new boron polyol, a slight increase in the density of produced PUR-PIR foams was observed in comparison to the reference foam. Foam density is in the range from 33.4 kg/m^3 for K0 foam up to 41.5 kg/m^3 for K5 foam, containing 0.5 R of the new polyol. The decrease in the apparent density of K1–K5 PUR-PIR foams is related to the amount of the boron introduced to the PUR system. By replacing the petrochemical polyol (with 9200 mPa s viscosity) with the boron polyol (with viscosity 283 mPa s), the total viscosity of the premix containing the new polyol was decreased. The amount of water used as a chemical foaming agent stayed at the same level for all recipes and generates the same amount of CO_2.

3.2. Influence of raw material type on thermo-insulating properties of obtained PUR-PIR foams

Rigid foams are mainly used as thermo-insulating materials. That is why a low heat conductivity of those materials is one of their most sought-after characteristic. The thermo-insulating properties of the obtained foam materials were determined by measuring the heat conductivity coefficient λ and the amount of closed cells (**Table 5**).

Rigid polyurethane-polyisocyanurate foams are cellular materials, where the amount of closed cells is significantly more prominent, and the share of open pores is in the range of

	K0	K1	K2	K3	K4	K5
Heat conductivity coefficient λ (mW/m K)	30.7	26.4	26.4	26.5	26.5	26.5
Amount of closed cells (%)	83.2	92.5	92.6	92.5	92.5	93.0

Table 5. Value of heat conductivity coefficient and amount of closed cells in obtained foams.

5–10%. At the same time, the amount of closed pores is dependent on their size. The gas closed in the foam cells and also the structure of a polyurethane matrix (but to a lesser degree), both take part in the heat flow process. The boron polyol used in the rigid PUR-PIR foam production causes an increase in the amount of closed cells in comparison to the K0 reference foam. The amount of closed cells in K0 foam is 83.2%. For foams containing tri[N,N′-di(methylenoxyethylentio-2-hydroxyethyl)urea], the amount of closed cells ranges from 92.9%, K1 foam, to 93%, K5 foam. It has been determined that the amount of new polyol does not influence the amount of closed cells; however, the application of those polyols during foam production has been significant. All rigid PUR-PIR foams obtained with the use of the new boron polyol have over 90% of closed cells. The increase in this number is not indifferent towards the value of the heat conductivity coefficient of those foams. In rigid foams designated for thermal insulation, the definite majority of cells should be closed so that the foaming gas could remain enclosed within the material. The value of the heat conductivity of the gas closed in cells comprises over 80% of the total value of heat conductivity coefficient of rigid polyurethanes. It is mostly dependent on the type of porophor, foam density and the amount of closed cells. It is worth keeping in mind that the thermo-insulating properties change during foam exploitation. It is with correlation to the changes in the composition of the gas mixture in the cells. By choosing the chemical method of foaming PUR-PIR foams, porous materials were produced that contained carbon dioxide in the cells. The polyurethane's cell walls are permeable for carbon dioxide which diffuses into the outside during material use due to its small particles. Then, oxygen takes place of the carbon dioxide. Foams that do not have a solid protective coating lose their initial insulating properties very quickly. **Figure 3** represents the changes in the values of heat conductivity coefficient for K0 and K5 foams in relation to the amount of days they were used.

It can be observed that the changes in the values of heat conductivity coefficient during the ageing process take longer time in the structures with the boron polyol. In the initial stage, the heat conductivity coefficient for the K0 reference foam was 30.7 mW/mK. During foam exploitation in room temperature, the λ-coefficient value increased to 37.2 mW/mK after 120 days. However, in the K5 foam, the initial λ-coefficient value was 26 mW/mK. During the ageing process in room temperature, the coefficient value increased to 28.9 mW/mK. The decreased heat flow in the foam modified with the new polyol is related to its more organized structure and larger amount of closed cells.

The heat conductivity coefficient depends mainly on the cell structure in the final product. That is why a thorough analysis of the cell structure is so important. At an early stage of the foaming process, there is a chain of reactions that start with the creation of micro-bubbles.

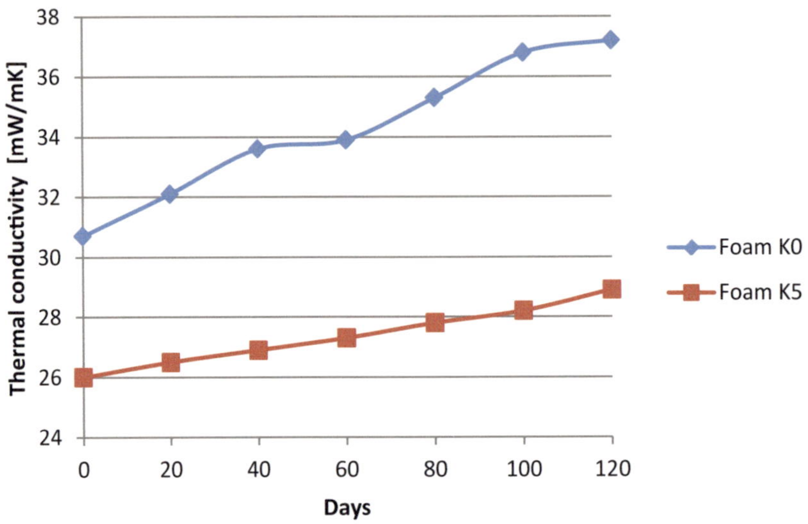

Figure 3. Changes in the values of heat conductivity coefficient for foam during ageing process in room temperature.

The cell growth comes next and is related to the porophor evaporation. Growing cells start to touch. Cell walls start to form in a form of thin membranes and ridges in the place where three cell walls meet. The construction stability is dependent on the chemical composition of the walls and ridges, and the size and shape of cells. The pictures of the obtained rigid foams using SEM method are depicted in **Figure 4**.

By analysing the SEM pictures of the cell structure, it can be observed that they depict internal closed-cell structure of the produced foams. The applied boron polyol supports the surface-active agent by creating more regular and smaller cells. By examining the shape of the closed cells in the photographs, it can be easily seen that the addition of boron polyol to the foam recipe contributes to a more uniform cell structure in comparison to the reference foam.

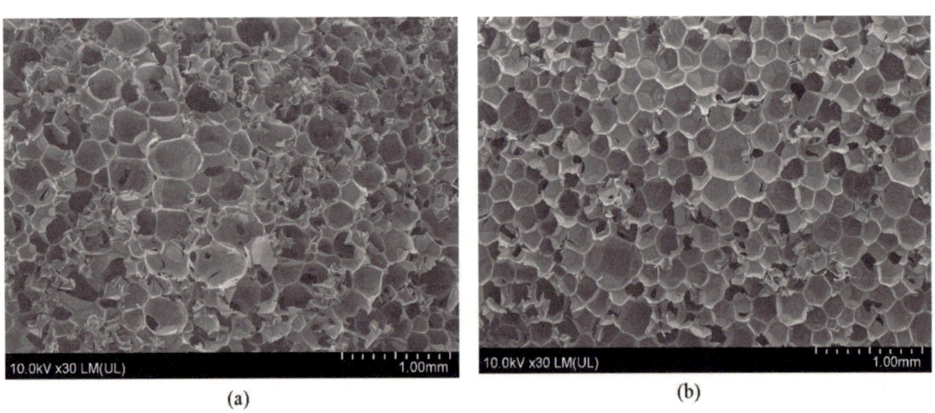

Figure 4. Cell structure of PUR-PIR foams: (a) K0 foam and (b) K5 foam.

3.3. Mechanical properties and dimensional stability of produced PUR-PIR foams

The structure of the polyurethane matrix has a significant influence on the mechanical properties and the durability of the cells in foam materials. By changing the recipe of the foams, some characteristics of the final product can be modified. However, the greatest significance in the shaping of PUR foam's mechanical properties is due to the apparent density and cell structure. The new boron polyol used in the foam composition improved the mechanical properties of the obtained rigid polyurethane-polyisocyanurate foams. **Table 6** presents the examination results of compressive strength, brittleness, water absorption and ageing.

The research conducted on mechanical properties of PUR-PIR foams helps determine that the new boron polyol significantly improves those properties. The compressive strength measured in a perpendicular direction increased after using the boron-nitrogen polyol in the foam production. The value increases along with the amount of the equivalent of the boron compound used in the foams. For the reference foam (without boron), the compressive strength is 238 kPa. For other foams, in which the petrochemical polyol Rokopolu RF-55 was subtracted to add the boron polyol, this value was in the range from 273 kPa (K1 foam with 0.1 R of borane tri[N,N'-di(methylenoxyethylentio-2-hydroxyethyl)urea]) to 339 kPa (K5 foam containing 0.5 R of borane tri[N,N'-di(methylenoxyethylentio-2-hydroxyethyl)urea]). The increase in compressive strength has to be correlated with the apparent density of the PUR-PIR foams (**Figure 5**).

The mechanical strength measured according to the foam's growth direction is usually higher than the one measured perpendicularly to the growth, because the cells are elongated in the direction in which the foam grows. The improvement in strength measured perpendicularly can be attributed to the changes in foam's morphology, that is, increased amount of polar urea groups capable of creating hydrogen bond. Additionally, the closed-cell structure gives the materials better compressive strength, in comparison to materials with large open cells.

The boron polyol used in the production of rigid polyurethane-polyisocyanurate foams reacts with the polyurethane composition in a way similar to a typical crosslinking compound, helping create a more unified foam structure. Because of that, the examination of brittleness

Foam	Compressive strength (kPa)	Brittleness (%)	Water absorption (%)	Mass loss (%)	Changes in linear dimensions (%)	Changes in volume (%)
K0	238	36.2	1.98	0.6	1.9	1.8
K1	273	11.9	0.85	0.4	1.8	1.4
K2	288	10.2	0.86	0.4	1.3	1.7
K3	295	9.6	0.86	0.9	1,5	1.5
K4	326	8.4	0.86	0.8	1.1	1.3
K5	339	8.0	0.85	0.6	1.6	1.7

Table 6. Mechanical and ageing properties of the rigid PUR-PIR foams.

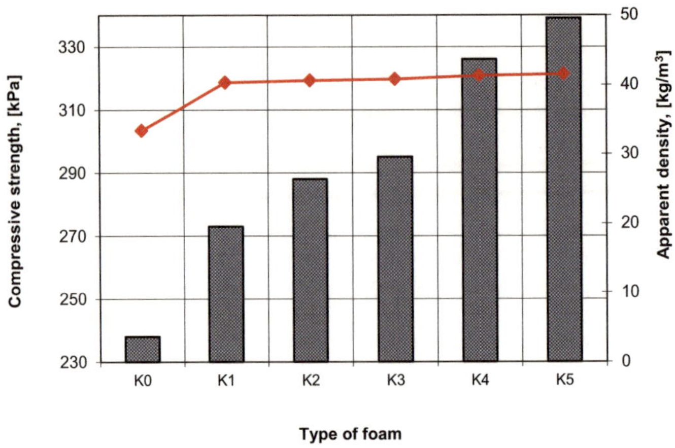

Figure 5. The correlation between compressive strength and apparent density, and the amount of boron polyol in the foams.

helped determine that the addition of the boron polyol into the foam recipe causes significant decrease in the foam's brittleness. The higher the amount of boron polyol in the foams, the lower the brittleness, from 36.2% for K0 reference foam to 8% for K5 foam modified with the new polyol. Similar to the compressive strength, the brittleness examination of the obtained foams shows a correlation between this value and the apparent density of the samples. Higher apparent density of PUR-PIR foams enables a significant decrease in their brittleness.

There is also a correlation between the compressive strength and the linear dimension stability. Along with the changes in the temperature, the internal pressure of the gas inside the cell changes. It creates a difference in pressure between foam cells and external atmospheric pressure. This difference of pressures needs to be lower than the foam's compressive strength to retain its dimensional stability. Foam deformation should not occur when the compressive strength is greater than 100 kPa, which is a value higher than the possible difference between atmospheric pressure and the pressure inside foam cells, which is close to zero. When comparing the results of the ageing test for the produced rigid PUR-PIR foams, it was determined that there is a strong correlation between the stability of linear dimensions, mass loss and changes in volume, and the use of the new polyol in foam recipe. During the simulated ageing of the samples, it was observed that mass loss did not exceed 1% for all foams. Similarly, the changes in linear dimensions did not exceed 2%. This qualifies those products for the use in thermal insulation.

3.4. Combustible properties of rigid PUR-PIR foams with new polyol

Rigid global statistical studies regarding fires in the recent years show that vast majority of fatal accidents during fires (60–80%) was caused by inhaling the products of thermal decomposition and burning, as well as limited visibility due to the generated smoke. The cause of a smoke with higher density and toxicity is incomplete burning of the gas phase. It also increases the ratio between carbon oxide and carbon dioxide, which makes the smoke more toxic. That is why the best way to lower the combustible properties of polyurethane materials is to produce them with raw materials that will help obtain a polymer with slow burning speed and low burning efficiency of flammable gas phase. Despite the obstruction of material combustion and limitation of flame spreading, the antipirene should not affect the processing and should not worsen the application properties of the product. The tendencies of boron and nitrogen atoms for supplementing and strengthening flame-retardant properties were used to determine the boron polyol. The percentage value of those atoms in the new polyol is around 14%; hence, positive results can be expected for the examination of non-halogen method of lowering the burning of new foams. Boron participates in endothermic reactions, whose end result is the release of water and creation of protective glasslike layer. This layer protects the base from oxygen and heat from the flame. Nitrogen compounds, on the other hand, decompose into gas products that take part in creating foamed carbon layers in the condensed phase. After reaching the gas layer, they burn and become free radical scavengers.

In order to test the foam behaviour in flame, various tests were conducted. Cone calorimeter method helps to define burning properties of the material and to characterize the reactions that take place during the burning process. The comparison of parameters of the pyrolysis process in selected rigid polyurethane-polyisocyanurate foams obtained with the use of new boron polyols has been presented in **Table 7**.

The produced rigid polyurethane-polyisocyanurate foams with the new polyol (K1 and K5), as well as the K0 reference foam (containing only the petrochemical polyol), were subjected to the burning process using cone calorimeter method. For the K0 reference foam, the combustion time was very short and equalled 1.48 s. The same time is characteristic for materials with porous structure that are highly flammable. After introducing the boron polyol to the foam recipe, the combustion time was significantly longer. For the K5 foam, containing the highest amount of the new polyol, the time was 14 s. On the other hand, in the K1 foam, containing the least amount of the polyol, the combustion time was shortened to 7 s. During the burning test using cone calorimeter method, the total heat release (THR) value was also measured, which determines the total heat released by the burning foam. The largest amount of heat, equalling 14.3 MJ/m^2, was released while testing the K0 reference foam. The use of the new polyol in the composition helped lowering the THR value, not exceeding 4 MJ/m^2. The total heat release value was the lowest for the K5 reference foam, with the largest amount of the boron polyol. The THR value can indicate that the new polyol shows cooling properties, by lowering this parameter by around 80%.

Foam symbol	Average time to steady combustion (T_{comb}) (s)	Total heat released (THR) (MJ/m$^{2)}$	Average mass loss rate (MLR) (gm^2/s)	Maximum heat release rate (HRR) (kW/m^2)	Time to reach max. HRR (s)	Amount of CO released (g/g)	Amount of CO_2 released (g/g)	Oxygen index (%)
K0	1.48	14.3	10.22	278.90	10	1.352	11.2	19.6
K1	7.00	3.7	7.03	221.21	27	0.589	1.6	23.2
K5	14.36	2.8	4.87	189.77	37	0.243	1.1	26.4

Table 7. Results of flame tests in selected rigid PUR-PIR foams.

The path of the heat release rate (HRR) curves shows information regarding the mechanism of lowering the flammability, which depicts the maximum value of the heat release rate. It is a very important parameter and an indicator of the material's tendency to self-extinguish in case of fire. The HRR values in **Table 8** were determined for data from the moment of combustion until the end of the test. **Figure 6** presents the curves showing the course of heat release (HRR) in the new foams.

The HRR curves of rigid PUR-PIR foams modified with the new polyol and the non-modified foam show subsequent stages of the burning process. It can be observed that initially the foams become warm, then the volatile parts and flammable gas products are produced. The burning of gases is the reason for creating large amounts of heat. The curves indicate that for the K0 reference foam, the rate of released heat is very energetic. Also, the flame on this foam is sustained. However, the foams modified with the new polyol burn in less rapid manner and reach lower values of heat release. The elongation of maximum HRR time value from 10 s (K0) to 37 s (K5) also shows flame-retardant properties of boron polyols.

The amount of exhausted CO is much higher for the reference foam than for the foams modified with the new polyol. Carbon oxide in the reference foam was emitted in the amount of 1.352 g/g. After using the boron polyol, the amount of carbon oxide was lowered almost four times, until reaching 0.243 g/g value for the K5 foam, containing 0.5 R of borane tri[N,N′-di(methylenoxyethylentio-2-hydroxyethyl)urea]. Similar correlation can be seen in the carbon dioxide emission during the burning test. The conducted analysis of carbon oxide and dioxide emissions uniformly indicates that the new boron polyol is an effective addition that reduces the amount of gases produced during a fire. Rigid polyurethane-polyisocyanurate foams with the new polyol have substantially longer time value for permanent combustion than the reference foam.

The oxygen index for the reference foam is 19.6%; however, in foams with borane polyol it is in the range from 23.2% (K1 foam) to 26.4% (K5 foam). By using the oxygen index method, it has been determined that the presence of the new polyol in the PUR-PIR foam helps reducing flammability of this material by abound 21%. During the examination in cone calorimeter, foams modified with the new polyol produced less charred residue. Also during this test, it was observed that the K5 reference foam did not burn but rather glowed. It is caused by minor mass loss of this foam (4.87 gm^2/s) during the burning test. The speed of fire spreading, flashover, is an important parameter used for the comparison of different materials with regard to fire safety. It is the value opposite to time to reach flashover ($1/t_{flashover}$) (Eq. (3)) [14]:

Foam sample	$1/t$ flashover, fire hazard (kW/m^2·s)	$RTFH_{CO/CO_2}$, relative toxic fire hazard
K0	188.4	2.8745
K1	31.6	0.1655
K5	13.2	0.0238

Table 8. Properties of selected rigid PUR-PIR foams determining their behaviour during fire.

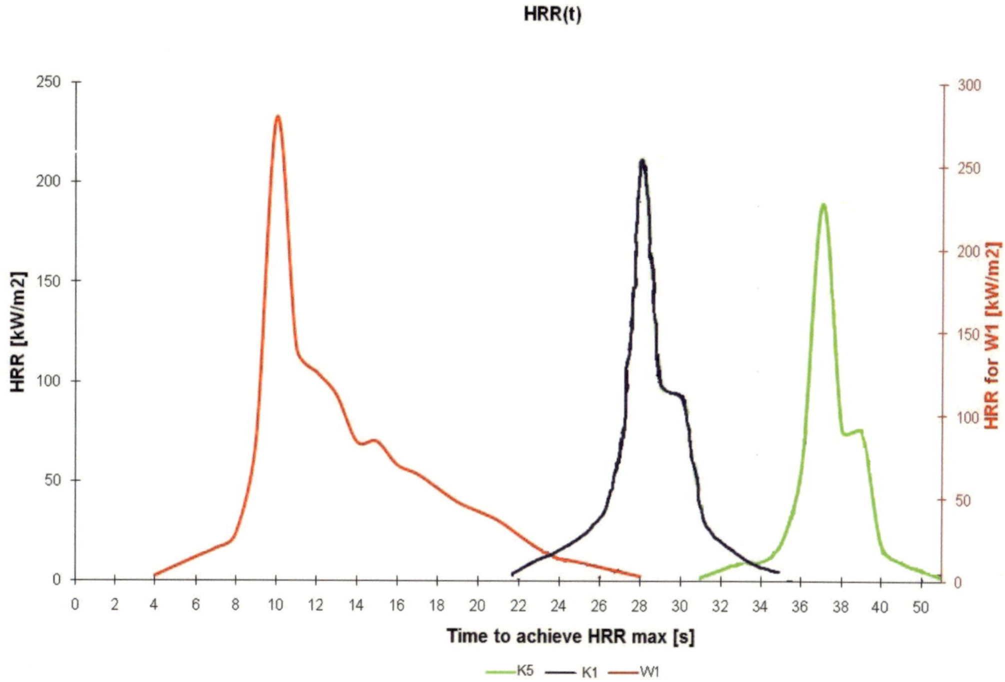

Figure 6. Heat release rate (HRR) for rigid PUR-PIR reference foam and foams containing new polyols (chart from cone calorimeter).

$$\frac{1}{t_{\text{flashover}}} = \frac{\text{HRR}_{\text{max.}}}{T_{\text{comb.}}} \tag{3}$$

where $\text{HRR}_{\text{max.}}$ is the maximum heat release rate; $T_{\text{comb.}}$ is the time to combustion.

Relative toxic fire hazard (RTFH) is another parameter that can help determine the fire danger rating. In this research, the determined RTFH indicators relate to carbon oxide and dioxide, because the use of cone calorimeter can measure only CO and CO_2 emissions. Those indicators are calculated according to Eq. (4):

$$\text{RTFH}_{\text{CO/CO}_2} = \frac{\text{MLR}}{\text{TTI}} \cdot \left(\frac{\text{CO yield}}{\text{LC}_{50}^{30}\ \text{CO}} + \frac{\text{CO}_2\ \text{yield}}{\text{LC}_{50}^{30}\text{CO}_2} \right) \tag{4}$$

where MLR is the average mass loss rate; CO yield is the average CO emission; CO_2 yield is the average CO_2 emission; LC_{50}^{30} CO and LC_{50}^{30} CO_2 are the lethal concentration of CO and CO_2, respectively, causing death of 50% of tested animals during 30-min exposition, in accordance with PN-88/B-02855.

The results of examined fire hazard related to the flashover and toxicity during the burning process of selected rigid PR-PIR foams are represented in **Table 8**.

The calculations show that the danger related to the speed of fire spread ($1/t_{\text{flashover}}$) in the tested rigid PUR-PIR foams can be lowered by using the new boron polyol. The value decreased from 188.4 $kW/m^2 \cdot s$ for the reference foam to around 13 $kW/m^2 \cdot s$ for the K5 foam, containing the largest amount of boron polyol. When looking at the values of carbon oxide and dioxide emissions, it can be stated that the use of the new boron polyol helps reducing the toxic fire hazard for the obtained rigid PUR-PIR foams. The toxic fire hazard for the K0 reference foam is 2.8745; however, the use of the new polyol in the foam production decreased this value by around 99%.

4. Summary

The examination results presented in this paper are the answer for the search of fire-safe polyurethanes with good application properties. The study proved that the modification of the premix for polyurethane foam production can give the foams the desired properties. The right selection of polyol materials, isocyanate and auxiliary compounds during the recipe creation is very important. Including the boron polyol in the foam recipe allowed obtaining new thermo-insulating material with good physicomechanical properties. The research showed that the new compound, which was a derivative of N,N′-di(methylenoxy-3-hydroxypropyle)urea and boric acid, could be used as a polyol in the production of polyurethane materials, and simultaneously act as flame-retardant agent. Complex study on the flammability conducted using various methods helped assessing the behaviour of produced rigid polyurethane-polyisocyanurate foams in case of fire. The research proved that the use of the new polyol, containing boron and nitrogen atoms, increased flame resistance. The efficiency of this resistance was mainly shown by the total value of released heat, prolongation of the time for steady combustion, lower amount of carbon oxide and dioxide emission, increase in the oxygen index and by the lower maximum burning temperature reached in shorter time in comparison to the reference foam.

Author details

Joanna Paciorek-Sadowska

Address all correspondence to: sadowska@ukw.edu.pl

Faculty of Technology, Kazimierz Wielki University, Bydgoszcz, Poland

References

[1] Król P. Polyurethanes – a review of 60 years of their synthesis and applications, Polymers, 2009;**4**(708):490.

[2] Randall D., Lee S., The polyurethanes Book, J. Wiley & Sons, Ltd. 2002, UK.

[3] Prociak A, Rokicki G, Ryszkowska J. New generation polyurethane materials, Wydawnictwo Naukowe PWN, 2014, 163-193.

[4] Szycher M. Szycher's Handbook of Polyurethanes. Boca Raton, USA: CRC Press; 2013

[5] Zatorski W., Brzozowski Z.K., Łebek K., Production of PUR and PUR-PIR foams with red phosphorus as a flame retardant, Polymers, 2005;**50**:686.

[6] McKenna S. T., Hull T. R., The fire toxicity of polyurethane foams, Fire Science Reviews, 2016;**5**:3. DOI 10.1186/s40038-016-0012-3

[7] Grand AF, Wilkie CA. Fire Retardancy of Polymeric Materials. M. Dekker New York, NY; 2000.

[8] Lemos VA, Santos MS, Santos ES, Santos MJS, dos Santos WNL, Souza AS, de Jesus DS, das Virgens CF, Carvalho MS, Oleszczuk N, Vale MGR, Welz B, Ferreira SLC. Application of polyurethane foam as a sorbent for trace metal pre-concentration - A review, Spectrochimica Acta Part B: Atomic Spectroscopy. January 2007;**62**(1):4-12.

[9] Chattopadhyay DK, Raju KVSN. Structural engineering of polyurethane coatings for high performance applications, Progress in Polymer Science. March 2007;**32**(3):352-418.

[10] Lubczak J., Chmiel-Szukiewicz E., Duliban J., Głowacz-Czerwonka D., Lubczak R., Łukasiewicz B., Zarzyka I., Łodyga A., Tyński P., Minda-Data D., Kozioł M., Majerczyk Z. Polyurethane foams with 1,3,5-triazine ring of improved thermal stability, Przem. Chem., 2014;**10**:1690-1697.

[11] Zhao C. Y., Yan Y., Hu Z. H., Li L. P. , Fan X. Z. Preparation and Performance Characterization of Polyisocyanurate(PIR)/Silica Aerogel Composite Rigid Foam Insulation Materials: Journal of Wuhan University of Technology, 2014;**36**(8):17-24.

[12] Feng X. Y., Li S. K., Wang Y., Wang Y. C., Liu J. X. The Effects of SiO2/PEG Suspension on Mechanical Properties of Rigid Polyurethane Foams, Advanced Materials Reserch, 2013;**815**:246-250.

[13] Lubczak J. Polyhydroxyalkyl derivatives and polyetherols obtained from azacyclic compounds. Part II. Reaction with formaldehyde and alkylene carbonates, Polymers. 2011;**56**(5):360-368.

[14] Trotzsch J. Plastics Flammability Handbook. 3rd ed. Munich: Carl Hanser Verlag; 2004

[15] Maurerem O. Organophosphorus compounds in respect of increasing fire safety requirements, Polyurethanes Magazine International. 2007;**04**:234.

[16] Paciorek-Sadowska J, Czupryński B. New compounds for production of polyurethane foams, Journal of Applied Polymer Science. 2006;**102**(6):5918.

[17] Czupryński B, Paciorek-Sadowska J, Liszkowska J. Modifications of the rigid polyurethane-polyisocyanurate foams. Journal of Applied Polymer Science. 2006;**100**:2020

[18] Lubczak J. Polyurethane foams with purine rings. Polymers. 2007;**52**:595-600

[19] Paciorek-Sadowska J. Modification of PUR-PIR foams by boroorganic compound prepared on the basis of di(hydroxymethyl)urea, Journal of Porous Materials. 2012; **19**(2):167-171

[20] Łukasiewicz B, Lubczak J. Oligoetherols and polyurethane foams with 1,3,5-triazine ring and boron atoms. Polymers. 2012;**57**:819-829

[21] Lubczak J. Reaction of uric acid with excess of propylene carbonate. Journal of Applied Polymer Science. 2006;**101**:2482-2487

[22] Kania E, Lubczak J. Polyurethane foams with pyrimidine rings. Polish Journal of Chemical Technology. 2014;**16**(3):1-6

[23] Lubczak J. Hydroxyalkylation of cyclic imides with oxiranes Part I. Kinetics of reaction in presence of triethylamine catalyst. Open Journal of Physical Chemistry. 2012;**2**(2):88-96

[24] Lubczak J. Hydroxyalkylation of cyclic imides with oxiranes. Part II. The mechanism of reaction in presence of triethylamine. Open Journal of Physical Chemistry. 2012;**2**(2):97-102

Chapter 6

Erosive and Abrasive Wear Resistance of Polyurethane Liners

Hossein Ashrafizadeh, Andre McDonald and
Pierre Mertiny

Additional information is available at the end of the chapter

http://dx.doi.org/10.5772/intechopen.68870

Abstract

Material removal caused by the impact and sliding of a stream of particles is a typical wear mode in the oil and gas industry. Protective coatings can be employed to increase the service life of equipment that is exposed to harsh erosive and abrasive environments. Among all the types of protective coatings and liners, polyurethane elastomers have received great attention owing to their excellent wear resistance and comparatively low cost that would allow for large-scale applications. The excellent wear resistance of polyurethane elastomers is a result of their high resilience and propensity to elastic deformation that enables the absorption of impact energy of erodant particles with minimal damage. The relation between the wear resistance of polyurethane and its mechanical properties has been the subject of previous studies. This chapter reviews the research that has been conducted to study the wear resistance of polyurethane elastomers. Testing apparatuses employed, material characterization techniques, evaluations of material removal mechanisms, and parameters with the strongest effect on wear resistance of polyurethane elastomers are herein explored. A review of finite element modelling approaches for in-depth study of the wear phenomenon of polyurethane elastomers is also presented in this chapter.

Keywords: finite element modelling, mechanical properties, polyurethane liner, wear resistance, wear testing

1. Introduction

1.1. Wear and corrosion in oil and gas industry

The advances in technology and higher demand for energy require enhanced rate of extraction and transportation of oil and natural gas fossil fuels. The use of pipelines for transporting of oil and gas products is a safer method compared to other possible ways of transportation as statistical analysis have shown in Ref. [1]. However, leaks and ruptures can still occur with significant hazard for humans [2]. Beside the life-threatening aspects, failures caused by degradation of pipelines can lead to serious economic losses [3] and environmental disasters such as pollution and contamination of wildlife and sea creatures [4]. Corrosion and wear in components and pipelines can be considered as one of the main causes of failure and leakage in oil and gas industry [5, 6]. All the three major fossil fuels, crude oil, and oil-sand and natural gas are very corrosive. Crude oil contains the corrosive ingredients such as carbon dioxide (CO_2), hydrogen sulfide (H_2S), organic acids, dissolved gases, and salt water [7], while the oil sand comprises of CO_2 and corrosive ions such as Cl^-, HCO_3^-, and SO_4^- [8]. Natural gas is also corrosive due to the presence of CO_2, H_2S, and some calcium and chlorine compounds [9]. On the other hand, in some sequences of extraction of the petroleum products, erodant particles such as sand may be mixed with the flowing fluid to form multiphase solid-liquid mixtures. The flowing of these mixtures in pipelines and equipment in oil and gas industry may lead to solid particle erosion in addition to corrosion. The combination of wear and corrosion can extensively reduce the lifetime of the equipment due to the higher rate of material loss [10].

Although the corrosion and wear of components exposed to harsh erosive-corrosive environments cannot be thoroughly eliminated, protective coatings can be employed to improve the life time of equipment and prevent early and unpredicted failures. The selection of the proper coating material depends on the service condition, financial aspects, and fabrication processes. In large-scale applications, such as transportation pipelines, polymer-based protective liners are preferred owing to their relatively low cost and ease in fabrication. Among the polymeric protective liners, polyurethane (PU) elastomers have received great attention due to their ease in processability, excellent resistance to corrosion, erosive and abrasive wear, and comparatively low cost that allows for large-scale applications [11, 12]. PUs are organic polymers with the urethane group in their chemical structure that can be synthesized by the reaction of a diisocyanate and a polyol [13]. Although the PU fabrication process is similar to the methods typically used for polymers, its mechanical properties such as high elongation at break and minimal plastic deformation are comparable to that of vulcanized rubber that has a more complicated production process [14]. PU has better wear resistance than most polymers [15], rubbers [16], stainless steels [15], and even some hard-faced tungsten carbide-cobalt (WC-Co) coatings [17]. Excellent resistance to wear and corrosion together with ease in fabrication process and low cost has made PU an excellent option for use as protective liners in large-scale applications such as oil and gas pipelines [18].

1.2. Abrasive and erosive wear

Abrasive and erosive wear are the two major wear mechanisms in conditions where the relative motion between the surface and hard erodant particles is responsible for damage and wear of the target material. In abrasive wear, the erodant particles are forced against the surface while moving along [13]. The erodant particles may slide or experience a combined sliding-rotating motion [19]. On the other hand, in erosive wear, the progressive loss of material occurs by the impact of hard particles that are moving in a gas or a liquid stream [13, 19]. The wear mechanisms of abrasive and erosive wear are discussed in the following.

In abrasive wear, the erodant particles are forced toward the target while sliding along the surface. As a result of this relative motion, small fragments can be detached from the surface by the cutting action of the sliding hard particles [20]. This wear mode that is one of the major types of abrasion is entitled as microcutting. The microcutting is usually the dominant mechanism of material removal in circumstances where the erodant particles are angular and harder than the target surface. Alongside with microcutting of the surface, the erodant particles may plough the surface by a combined action of cutting and plastic deformation to form groove shaped defects on the surface. The two wear modes of microcutting and ploughing are categorized as cutting mechanisms. On the other hand, in conditions where the grit media are blunt, the accumulation of residual strains together with fatigue mechanism caused by the repeated deformation of the surface is the major mechanism of material removal from the surface. The cracks formed from material defects will propagate by the repeated loading-unloading, leading to reduced strength of surface and loss of material. **Figure 1** shows a schematic of the material removal from a ductile surface caused by cutting and plastic deformation in abrasive wear. In hard brittle surfaces such as ceramics, the fracture of the surface, crack formation, and detachment of small pieces as a result of crack intersections is considered as the major abrasive wear mechanism.

The mechanism of material removal in erosive wear is not only a function of the properties of the target surface but also a function of the testing condition such as velocity and impact angle

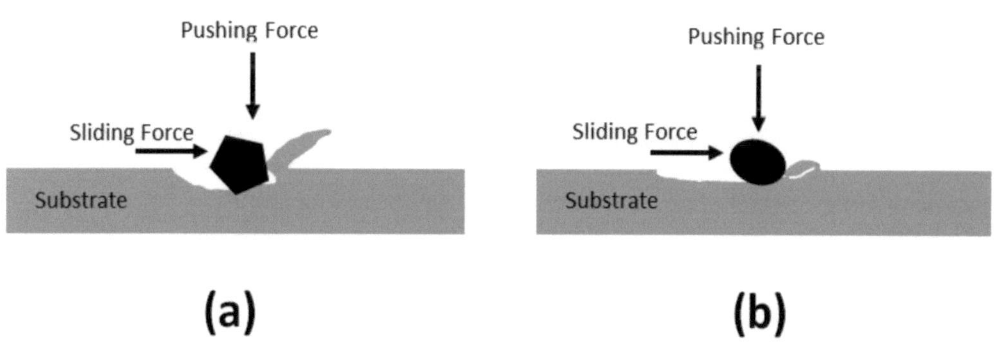

Figure 1. Abrasive wear of ductile substrates: (a) cutting mechanism and (b) plastic deformation together with fatigue mechanism.

of the erodant particles [21]. In conditions where the erodant particles impact the surface at low angles with respect to the surface (10 to 30°), the cutting mechanism similar to abrasion will be dominant. At low angles, the particle's normal impact force is high enough to enforce the particle for partial penetration, while the tangential force slides the particle along the surface to microcut small pieces from the target surface. As the impact angle increases, the tangential force produced upon impact will not be high enough to cut pieces from the surface. Alternatively, at higher impact angles of 60–90°, ductile targets will mostly experience plastic deformation, and the material removal occurs due to the microforging and extensive deformation of the target surface. The chips formed will detach from the surface at subsequent impacts due to the further accumulation of the residual strains and final detachment of the formed ridges [20]. Since the chipping mechanism of material removal requires higher number of impacts compared to cutting mechanism at oblique impacts, the ductile substrates present a minimal erosion rate at high impact angles. In contrast, the brittle ceramics have the highest erosion rate at normal impact angles, since the particle's normal force is maximum leading to higher fracture, cracking, and damage of the surface of the brittle substrate.

In many engineering applications, such as slurry motion of particles in a pipe, the material removal mechanism can be considered as a combination of erosive and abrasive wear. The slurry particles flowing in a pipe may slide onto the pipe bottom surface while being pressed toward the pipe surface by gravity and fluid weight. The motion induced by the flowing fluid and the pushing force can lead to abrasive wear of pipe material. On the other hand, the particles that are freely moving along the fluid stream and suddenly impacting the surface due to the flow turbulences within the pipe represent the erosive wear at low impact angles. Slurry flow in an elbow can be mentioned as another example for conditions in which combined abrasive-erosive wear may occur. Some particles are sliding while being pushed toward the surface, whereas some other particles are freely impacting the surface due to their kinetic energy and inertial forces. Consequently, when studying the wear of protective coatings and liners, the resistance of material versus both abrasive and erosive wear should be evaluated.

2. Wear of polyurethane liners

2.1. Material and mechanical characterization methods

The resistance of PU elastomers to abrasive and erosive wear is a function of both the testing condition and the properties of the PU. Ductility or softness, yield stress, elasticity, elastoplastic, and viscoelastic behavior are all parameters that can affect the stresses produced within the PU upon impact of erodant particles and, therefore, its wear resistance. To that end, when studying the wear behavior of PU elastomers, identification of the properties with the most significant effect on wear resistance of PU is of great interest. In this section, common experimental techniques to study the mechanical properties of PU elastomers are discussed. Through different testing techniques available for the characterization of PU, focus will be given to testing procedures for determination of parameters that correlate with the abrasive and erosive wear of PU.

2.1.1. Hardness testing

The hardness testing can be employed as a first-order approximation of ductility of elastomers and plastics [11]. The hardness of elastomers can be measured by a durometer according to the ASTM Standard D2240 testing practice [22]. In this method, the hardness of an elastomer is measured based on the penetration depth of an indenter into the substrate surface. The indentation depth is a function of elastic modulus and viscoelastic behavior of the elastomer [22]. The simplicity of this technique is one of its greatest advantages that allows for quick laboratory testing and in-field evaluations. Although determination of the exact value of elastic modulus is not feasible by this testing technique, the values measured are excellent for comparative evaluation of the elastomer softness. Moreover, monitoring of the indentation depth of the indenter with time can provide data about the viscoelastic response of material and its creep behavior. The hardness testing by durometer according to ASTM Standard D2240 does not provide any information about the elastoplastic behavior.

In a few previous studies [23, 24], it has been shown that the Vickers micro hardness testing that is usually employed for evaluation of the hardness of metals and ceramics [23, 25] can be used as a tool to provide information about the viscoelastic and elastoplastic behavior of polymers. Due to the high elastic deformability of elastomers, the shape of the indent is different from that of metals and a perfect symmetric indentation will not form on the surface. A variation in dwell time of the indenter during the test and also monitoring of the changes in size of the formed indents by time can provide information about the viscoelastic response of elastomers.

2.1.2. Tensile and compressive testing

Data obtained by durometer or Vickers hardness testing do not provide detailed information about the elastomer properties such as the Young's modulus and yield strength. Thorough characterization of PU elastomers can be achieved by testing equipment capable of tensile and/or compression testing at controlled load and displacement. Tensile tests conducted up to failure of the sample can provide information about the elastic modulus, final strength, and elongation at break of the polymer. These are the parameters that can affect the final resistance of a polymeric material to abrasive and erosive wear [21]. The sample size and the applied load and displacement can either be selected from standard testing practices such as ASTM Standard D638 [26] or may be selected similar to the type of loading that occurs in actual erosion or abrasion processes. Tensile testing can also be employed to study the effect of strain rate on the stiffness and strength of materials by testing at different strain rates. This is extremely important when studying the erosive wear of polymeric elastomers since the impact occurs within microseconds at very high strain rates [27]. Beside the tensile testing at different strain rates that can provide information about the viscoelastic response of polymeric materials, relaxation and creep testing can be conducted by this type of testing apparatus either in the form of tensile or compressive stresses. In relaxation tests, the stress relaxation of a material is studied after a sudden displacement at the beginning, while in creep testing, the deformation of the material is monitored upon exposure to a certain stress value [28]. Furthermore, tensile tests can be conducted at controlled temperature to evaluate the effect of temperature on the mechanical response of the studied PU elastomers.

The elastoplastic behavior of PU can be determined by cyclic loading in the form of tensile or compression testing. Cyclic loading-unloading can also provide information about the stress softening (Mullins damage) of elastomers, which is a permanent nonreversible damage to the structure of the material caused by loading [28]. Information about the elastoplastic response of elastomer is essential when studying the wear behavior since in abrasive and erosive wear, the repeated impact of erodant particles produces repeated loading-unloading on the elastomer surface [21].

2.1.3. Rebound resilience

The PU elastomers have better erosion resistance than most metals owing to their softness and high capability for elastic deformation. In fact, the PU elastic deformation enables absorbing the kinetic energy and gradual decelerating of the impacting particles with minimal damage. The kinetic energy absorbed in the form of elastic strain energy will be released later to rebound the erodant particle from the surface. The rebound resilience of PU can be employed as a parameter representing the ability of the elastomer to absorb kinetic energy of the erodant particle upon impact. This property can be measured according to the ASTM Standard D2632 [29]. In this testing practice, a plunger is dropped on the top of the sample surface from certain height. By recoding the rebound height of the plunger, the energy lost during the impact can be calculated. In a fully elastic deformation of the surface upon impact, the plunger would rebound to its initial height. Substrates with higher plastic deformation will restore smaller amounts of plunger energy and the plunger will rebound to a reduced height.

2.1.4. Dynamic mechanical analysis

The viscoelastic and time-dependent behavior of polymers can be characterized by either tensile/compression testing at controlled strain rates or by dynamic mechanical analysis (DMA). DMA can be conducted with different sample geometries, and since the vibration amplitudes by DMA are usually less than 100 μm, the DMA electronic shaker can vibrate the samples at very high frequencies [30]. Thus, the mechanical response of the studied PU can be categorized at different vibration frequencies or strain rates. Due to the relative small vibration amplitude in DMA, the measured properties are usually within the elastic region of the elastomer response. The data obtained from DMA are reported in the format of storage and loss modulus [30].

2.1.5. Thermal properties

The thermal properties of PU elastomers can also affect their resistance to erosive and abrasive wear. Heat can be generated during abrasion and erosion of PU elastomers by two mechanisms of (a) hysteresis and repeated deformation of PU and (b) the friction forces between the erodant particles and the target surface. The thermal properties of PU elastomers such as thermal conductivity and heat capacity can affect the temperature distribution below the impacted surface and, therefore, the wear resistance of the elastomer. Thermal conductivity of polymers can be measured by thermal constant analyzers [31]. Furthermore, the heat- and thermal-based test procedures such as differential scanning calorimetry (DSC) can provide information about the chemical structure of the elastomer such as glass transition temperature and melting points of the elastomer [28]. This information enables the selection of protective liners to ensure that the elastomer will remain within its rubbery phase in operation.

2.1.6. Testing procedures for evaluating abrasive and erosive wear resistance

The resistance to abrasive and erosive wear can be determined by conducting standard wear testing procedures. In most of the wear testing practices, the volume loss for a specific period of time will be employed as the parameter representing the wear resistance of the material. The volume loss can be measured either by optical techniques or simply by measuring the mass loss and further calculating the volume loss according to the density of the tested material. Clearly, larger volume loss for a given time is representative of lower resistance to wear. From among the different types of available abrasion testing procedures, the ASTM Standard G75, B611, and G65 are the most commonly employed testing procedures to study the abrasive wear of materials [32–34]. The ASTM G75 is the standard testing practice for determination of resistance to abrasion caused by slurry [32]. In this testing condition, the samples move in reciprocating motion while being forced toward a surface covered with slurry. While in this testing condition, wear occurs at both forward and backward motion, in ASTM G65 and ASTM B611, a rotating wheel is used to abrade the test coupons in a single direction. ASTM G65 is the standard test method for measuring abrasion resistance using a dry sand/rubber wheel apparatus [34]. **Figure 2** shows a schematic of the ASTM G65 abrasion testing procedure. Although this test has been widely used in previous works for evaluating the abrasion resistance of metals and ceramics, the heat produced in this test caused by friction forces can lead to unreliable results when evaluating the wear resistance of heat-sensitive substrates such as PU. ASTM B611 is very similar to ASTM G65, though the test is conducted in slurry, and the rotating wheel is made of steel rather than rubber [33]. This testing method may be preferable over ASTM G65 since the wet slurry can cool the sample during the test. It should be noted that in abrasion tests a wet area is in contact with the PU, and care should be taken to ensure that the wet environment will not affect the wear resistance by a possible chemical reaction between the PU surface and the wet slurry, and also a change in properties of PU by water absorption.

ASTM Standard G76 can be mentioned as the most commonly used standard test for evaluating the erosion resistance of different target materials [35]. In this testing scheme, the erodant particles are accelerated in a gas jet prior to impacting the surface at a desired angle. **Figure 3** shows a schematic of this testing procedure. This testing technique is advantageous for erosion testing of heat-sensitive substrates such as PU since the high velocity of the impacting jet can mitigate the adverse effects of a temperature rise during testing.

2.2. Relation between mechanical properties and wear resistance of polyurethane

The superior wear resistance of PU is due to its softness and high capacity for elastic deformation. In fact, the PU softness and high elastic deformability allow for gradual deceleration of the erodant particles while absorbing their kinetic energy. The kinetic energy absorbed will be released later to rebound the erodant particle from the surface. Accordingly, a study of the

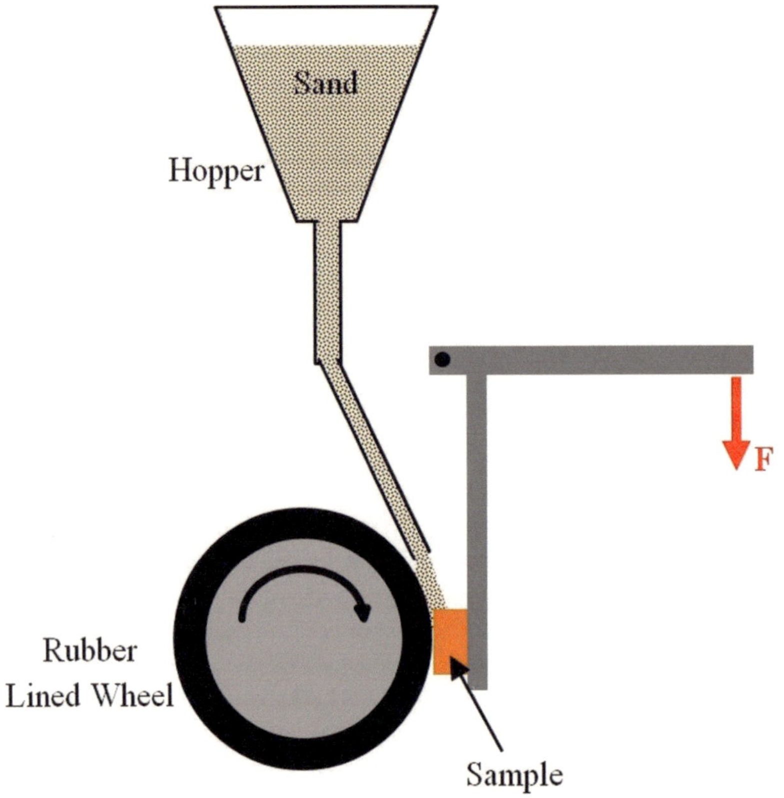

Figure 2. Schematic of standard testing method to measure abrasion resistance by dry sand/rubber wheel apparatus [34].

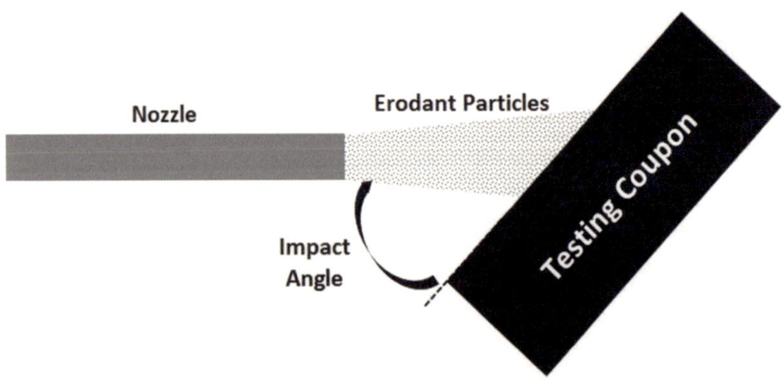

Figure 3. Schematic of erosion testing procedure according to the ASTM Standard G76 [35].

effect of PU softness on the wear resistance of PU has been the subject of a number of previous studies. The PU hardness as measured by durometer or Vickers hardness testing can be a representative of material softness and overall ductility as was discussed in Section 2.1.1 of this chapter. Li et al. [11] studied the erosion resistance of a series of castable PU elastomers with hardness values ranging from 20 to 90 IRHD. A trend of an increase in erosion rate with increasing hardness was observed. It was further shown in this study that the erosion rate does not correlate with the elongation at break of the studied polymeric elastomers. The effect of softness on the erosion resistance of rubber elastomers has also been the subject of previous studies. In a study by Zuev et al. [36], the effect of slurry temperature on the erosion rate of rubber elastomers was studied. It was found that the erosion rate decreased when the temperature was increased from 20 to 70°C. Since rubber would become softer at elevated temperatures, it can be concluded that the increase in softness at higher temperatures could have been the reason for the reduction in the erosion rate. Similarly, in another study by Marei et al. [37] where the erosion resistance of rubber was evaluated at elevated air temperatures, it was reported that the softer elastomer had lower erosion rate. At elevated temperatures, the rubber became softer due to the greater difference between the testing temperature and the rubber glass transition temperature [37]. Consequently, the softness is certainly a factor affecting the erosion resistance of elastomers, including PU. The softness and high deformability of elastomers enable the deceleration of the impacting particles at a longer time compared to hard surfaces such as metals or ceramics. As the impact time becomes longer, the impact forces and the stresses decrease accordingly due to the impulse formula. The impulse formula states that the force is the time derivative of momentum as:

$$m\vec{v} = \int \vec{F}_{\text{impact}}\, dt \tag{1}$$

where m is the particle mass, Δv is variation in particle velocity vector, F_{impact} is the impact force, and t represents time. Thus, the impact stresses are smaller in softer materials due to the longer impact duration.

Evaluating the effects of hardness on the resistance of PU to abrasive wear has also been the subject of previous studies. Hill et al. [38] evaluated the wear performance of PU by employing an abrasion testing procedure according to the ASTM Standard G65 [34]. The results from this study supported the validity of the graph of wear rate versus hardness proposed by Pitman [39] as shown in **Figure 4**. As seen in **Figure 4**, the abrasion resistance of PU does not vary significantly with hardness in Region B. The graph in **Figure 4** also shows that reducing the hardness of PU to very low values, Region A, increases the wear rate. This behavior is in contrast with erosion of PU elastomers in which the softer PU elastomer has higher resistance to erosive wear. The graph in **Figure 4** suggests that except for very hard PU elastomers (Region C), the harder PU elastomers have better resistance to abrasive wear, since the harder PU can better resist the penetration of erodant particles during the abrasion testing. The reduced penetration means smaller stress level and, therefore, reduced damage to the substrate caused by repeated deformation of PU.

Although the PU hardness seems to correlate well with the resistance to abrasive and erosive wear, PU elastomers with similar hardness values may have different resistance to abrasive and erosive wear [21, 40]. Ping et al. [17] evaluated the erosion resistance of two PU samples with relatively similar hardness values. In this study, the elongation at break obtained through tensile testing was introduced as a parameter that can affect the wear resistance of PU elastomers. It was shown that PU with higher elongation at break (320%) had higher resistance to erosive wear compared to PU with lower elongation at break (250%). Zhang et al. [16] also showed that

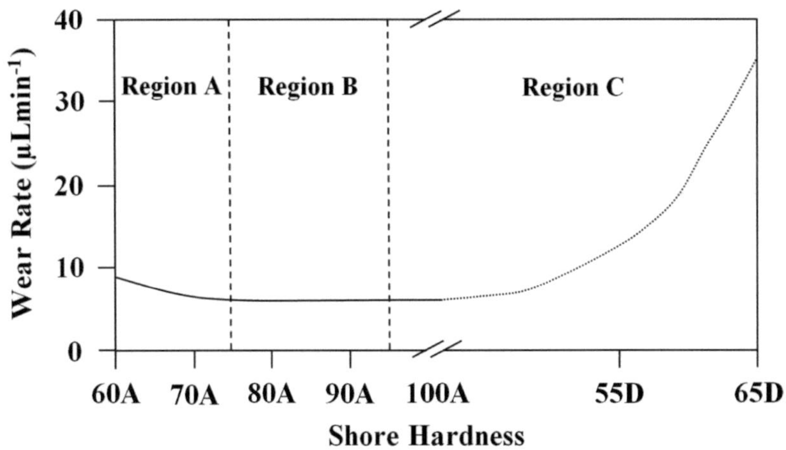

Figure 4. Abrasive wear rate of PU as a function of hardness [38, 39].

PU elastomers with higher elongation at break (520%) had higher resistance to erosive wear compared to other polymers with lower elongation at break such as polytetrafluoroethylene (150%). Similarly, Ashrafizadeh et al. [21] showed that a reduction of the elongation at break of PU at 100°C led to a sudden increase in erosion rate of PU at that temperature. According to the impulse formula, the softer material enables longer impact time and, therefore, reduced stresses and damage. However, the elongation at break of the material should be high enough to enable the deformation of the soft elastomer without failure. In fact, if the deformation strain caused by the impact of the erodant particle exceeds the strain at break, failure of the material will occur leading to detachment of fragments from the surface.

Tensile cyclic loading of PU samples allows for comparing wear resistance, hysteresis and elastoplastic behavior. Hysteresis of a polymer represents the fractional energy lost in a deformation cycle [21]. Beck et al. [41] conducted cyclic loadings to study the effect of PU hysteresis on wear resistance. It was found that PU elastomers with similar hardness values had different wear rates due to the differences in hysteresis of the studied elastomers. PU elastomers with higher hysteresis exhibited higher erosion rate. Larger hysteresis can negatively affect the strength of an elastomer in two ways: (a) higher heat production and temperature rise below the worn surface and (b) greater permanent irreversible damage to the polymer structure upon loading [28]. Thus, a material with a higher hysteresis not only suffers from adverse effects of temperature rise but also experiences a higher damage level upon impact of erodant particles. This can accelerate the progressive damage caused by the repeated impact of particles leading to final removal of material from the surface at a higher rate.

The relation between the elastoplastic behavior of elastomers and their wear resistance has been the subject of few previous studies. In a recent study by Ashrafizadeh et al. [21, 42], the elastoplastic response of PU elastomers obtained by cyclic tensile loading was compared with the data obtained from erosion testing at controlled temperatures. The results obtained showed that PU elastomers with higher residual strain (permanent set) upon unloading exhibited a higher erosion rate. This behavior was due to the fact that in PU elastomers with lower residual strain, a higher number of impacts will be needed for progressive damage and final detachment of fragments from the surface. On the other hand, in some studies, the elastoplastic behavior of elastomers has been assessed by evaluating the rebound resilience of the elastomer. For example, in a study by Hutchings et al. [40], rebound resilience was found to be the most dominant factor affecting the wear resistance of rubber elastomers in which the rubber with higher rebound resilience had the highest erosion resistance. It should be noted that measuring the rebound resilience is an approximation of the elastoplastic response behavior since this test only provides information about the elastoplastic response for a single loading condition related to the mass and velocity of a falling plunger [29].

In an assessment of the wear resistance of elastomers, attention should be given to the possible effect of temperature and chemical reactions on the overall wear resistance of PU protective liners. The mechanical properties of PU are sensitive to temperature and may vary significantly even by changing the temperature by around 40°C [21]. Thus, the temperature rise during the wear experiment may affect the erosion resistance of PU [14, 16, 38, 41, 43–46]. This suggests that accurate monitoring of the temperature during wear testing of elastomers is required. A review of the effect of temperature on wear resistance of elastomer is discussed in Section 2.4 of this chapter. On the other hand, the fluid that is in contact with the elastomer surface either as a cooling agent in abrasion testing or as a jet for accelerating the erodant particles in erosion testing can affect the wear resistance of elastomers in two ways. First, a chemical reaction may take place between the flowing fluid and the elastomer surface. For example, it was shown by Zuev et al. [36] that the erosion resistance of rubber is a function of the resistance of the elastomer to chemical reaction with the aggressive media. It was shown that as the concertation of acetic acid increased in the abrasive medium, the wear rate increased. Second, swelling of PU elastomer can affect the resistance to wear. Siegmann et al. [47] showed that for conditions in which the PU was in contact with organic fluid medium, the PU became softer by absorbing the solvent. It was found that higher swelling of the PU led to higher abrasive wear due to the softening of the PU. Thus, testing in dry conditions and controlled temperatures may be ideal when wear testing of PU elastomers to minimize the possibility of swelling and chemical reactions.

2.3. Mechanism of material removal

The mechanism of material removal in abrasion and erosion of PU elastomers by solid particles is a function of the wear testing procedure, properties of abrasive media, and mechanical properties of the target material. The wear mechanisms that have been suggested for PU elastomers can be divided into three categories of (a) cracking below the worn surface, (b) formation and detachment of ridges and (c) random scratches and gouges.

2.3.1. Cracking below the worn surface

In erosive and abrasive wear, compressive and shear stresses are produced by the impact or sliding of the erodant particle [48]. Due to the deformation of the elastomer in front of the surface, the produced stresses in that area are compressive. On the other hand, the stresses formed by friction forces and dragging of the elastomer are mostly shear as demonstrated schematically in **Figure 5**. The shear stresses generated by the friction forces have a maximum value at a certain depth below the surface (see **Figure 5**) [48]. The elastomer will be retrained from cracking near the surface since the compressive stress has the maximum value. However, as the distance from the surface increases, the compressive stress decays faster than strain does and, thus, depending on the wear process parameters and the fatigue properties of the elastomer, at some depth below the worn surface layer, the shear stresses produced by the repeated impact of particles will lead to crack formation and further propagation onto the PU [48]. The typical cracks formed at a certain depth below the surface have been observed in images taken by scanning electron microscopy (SEM) from the cross sections of a worn PU as shown in **Figure 6**. The existence of a maximum value for the shear stress at a certain depth from the surface has been also shown quantitatively in previous studies by finite element modelling [14]. Detachment of fragments and, therefore, wear of the target material occur as a result of the intersection and extension of cracks formed below the worn surface [14, 16, 43, 48]. The surface of PU samples worn by this mechanism does not have regular patterns. Cracks and detached pieces are scattered throughout the worn surface.

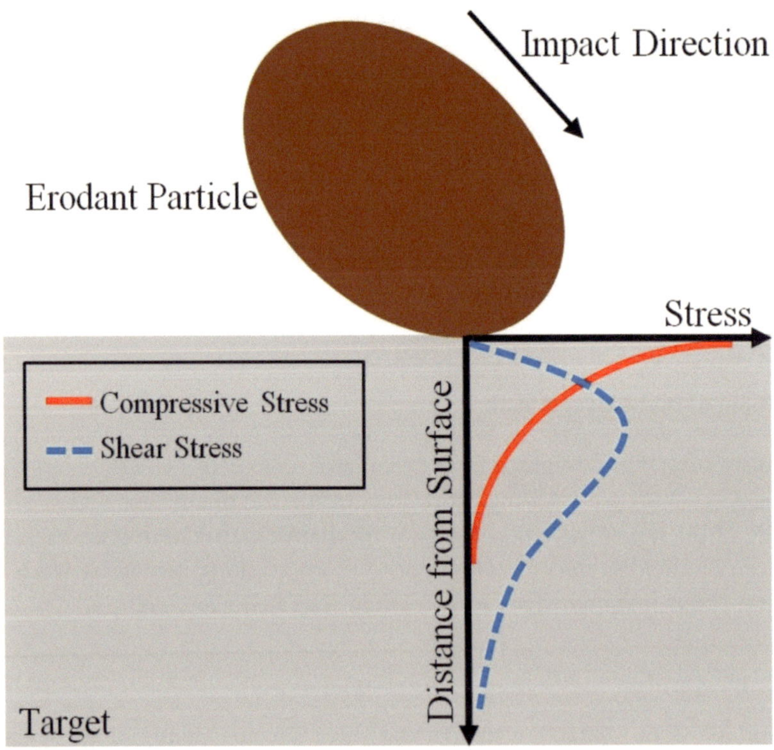

Figure 5. Schematic of stress distribution during abrasion of polymeric elastomers [48].

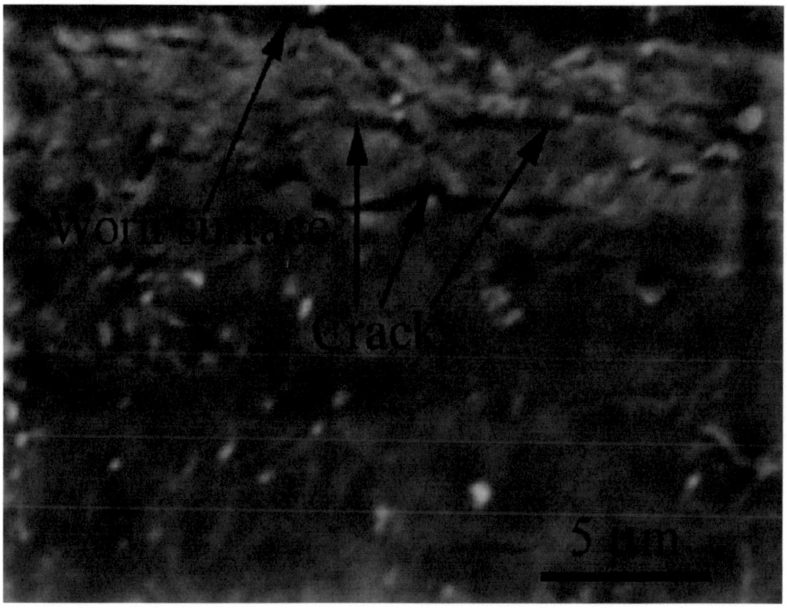

Figure 6. Cracks formed below the surface during abrasion of PU elastomer [48].

2.3.2. *Formation and detachment of ridges*

For conditions in which the stresses produced by the impact or the sliding of hard particles are smaller than the final strength, but higher than the yield strength of the PU elastomer, no loss of material will occur by a single impact. Alternatively, the gradual plastic deformation and formation of ridges on the worn surface are responsible for material loss from the surface. In this mechanism, as a result of repeated impact or sliding of solid particles, the plastic strains will accumulate to generate localized ridges on the PU surface. The ridges formed are perpendicular to the direction of impact or sliding of erodant particles [11]. In other words, the single impact of an erodant particle does not lead to material loss from the surface, and many successive impacts are required prior to damage and material loss from the surface in this mechanism [12]. **Figure 7** shows a typical image of such ridges formed after erosion testing of a PU elastomer, while **Figure 8** shows a side magnified view of one of the ridges. In these figures, the arrow shows the impact direction. The further accumulation of plastic strains will eventually lead to cracking at the bottom of the asperities, followed by the final detachment of the material from the surface. A typical crack produced on the base of one of these ridges is indicated in **Figure 8** by a circle. Consequently, this wear mechanism highly depends on the elastoplastic behavior of PU; PU elastomers with a higher tendency to revert to its initial condition upon loading and minimal plastic deformation have higher resistance to erosive wear since a higher number of impacts will be required to form and detach these ridges from the surface.

The formation of ridges perpendicular to the direction of impact and final fracture of the asperities is the mechanism of material removal in both erosive and abrasive wear of PU elastomers [11, 12, 14, 21, 38, 41, 46, 48]. The morphology and distances of the asperities produced

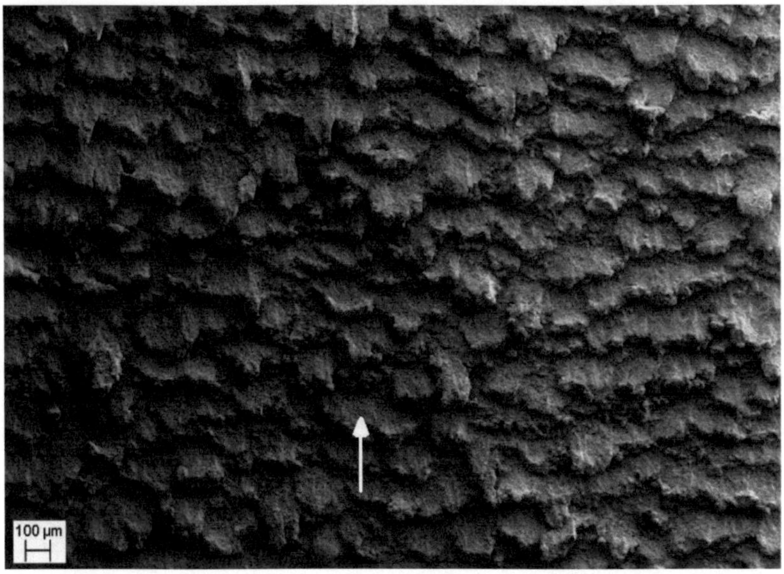

Figure 7. SEM image taken from the top surface of an eroded PU [21].

Figure 8. Side SEM image taken from one of the ridges formed on the surface of an eroded PU [21].

on elastomer surfaces are functions of the mechanical properties of the target surface. For example, it has been shown by Hutchings et al. [40] that ridges are more regular in shape and pronounced in rubber with higher resilience. Ashrafizadeh et al. [21] showed that the ridges formed were smaller for PU elastomers with lower elongation at break. The patterned ridges formed during abrasion of PU elastomers are also a function of the mechanical properties of the elastomer as was shown by Hill et al. [38], that is, asperities were closer in harder PU elastomers. Although a few studies have focused on evaluating the effect of elastomer mechanical properties on the shape of the ridges produced, further research for an in-depth understanding about the relation between the mechanical properties of the tested elastomers, testing condition, and the shape of the asperities formed is required.

2.3.3. *Random scratches and gouges*

In abrasive and erosive wear of PU elastomers, the material detachment may occur even by the impact of a single erodant particle based on the testing parameters and the properties of the elastomer. This mechanism of material removal is more similar to erosive and abrasive wear of metals in which the primary mechanism of material removal is random scratches and gouges on worn surfaces due to the cutting and gouging action by angular grit media. This type of wear usually occurs when the erodant particles have sharp edges to tear the elastomer surface. In other words, in conditions where sharpness or high velocity of erodant particles lead to production of stresses higher than the final strength of the elastomer, detachment of small fragment from the surface can occur. It should be noted that there is no regular pattern on the surfaces of elastomers worn by this mechanism. The worn surface by this mechanism is covered with cracks and detached fragments similar to the fatigue crack growth mechanism that was explained in Section 2.3.1.

2.4. Effect of working temperature on wear resistance of PU

The effect of temperature on wear resistance of PU elastomers has been of attention due to the heat sensitivity of PU and the possibility of heat production and, therefore, temperature rise during the abrasion and erosion processes. Unlike metals, the mechanical properties of PU may vary significantly even by temperature variation within the range of ±40°C. The fact that the temperature rise of as much as 50°C can occur during the wear process of PU elastomers emphasizes the importance of knowledge about the effect of temperature on the wear resistance of PU [38, 43]. In this section, the effect of temperature on wear resistance of PU and rubber elastomers will be discussed and previous studies related to this topic will be reviewed.

The heat generated during the wear testing of elastomers can be employed as a heat source to evaluate the effect of temperature on the wear resistance of PU elastomers. Hill et al. [38] evaluated the wear performance of PU by employing an abrasion testing procedure based on the ASTM Standard G65 [34]. Two testing procedures were conducted: (a) continuous abrasion testing and (b) abrasion testing with 10 min rest periods every 1.5 min to allow for cooling of the samples. It was found that the samples that were tested with the continuous process (procedure (a)) did not have a constant wear rate due to the uniform temperature within the samples during the test. It was found that the temperature rise affected the wear rate by varying the hardness of the PU. Zhang et al. [43] employed a grit blasting chamber to evaluate the effect of thickness of a PU liner on its erosion performance. It was found that the heat generated by hysteresis and friction forces increased the temperature of the PU in the layer beneath the surface. The increase in temperature negatively affected the strength of the PU material leading to lower erosion resistance. Even though, the effect of temperature on abrasive and erosive wear of PU elastomers was addressed, in these studies no external heat source for accurate and uniform control of the temperature during wear testing was employed. Accordingly, in some previous studies, testing assemblies capable of erosion testing at controlled temperatures by employing an external heat source has been developed. Zuev et al. [36] conducted erosion testing at elevated temperatures by controlling the slurry temperature. The increase of slurry temperature from 20 to 70°C improved the erosion resistance of the rubber owing to the improvement in elasticity and softness of rubber at the elevated temperature of 70°C. Marei et al. [37] also reported improvement in erosion resistance of rubber at elevated temperatures. In this study, an air blasting test scheme with the controlled temperature on the input gas was developed. It was found that at testing temperatures with greater difference from the glass transition temperature, the erosion rate of rubber was lower. In a more recent study by Ashrafizadeh et al. [21, 42], a test assembly for erosion testing at controlled temperatures was designed and developed. A cold spray system with controlled gas temperature and temperature controller and cartridge heaters were employed to heat the samples from the exposed and unexposed surfaces, respectively. The accurate temperature field within the samples during the erosion testing was further determined by a finite element numerical heat transfer model. In this study, the effect of temperature on strength, elongation at break, and elastoplastic behavior of PU elastomers was also studied and compared with their wear resistance. This comprehensive study showed that the increase in temperature may improve the erosive resistance of PU elastomers in two ways. First, the increase in softness of PU at elevated temperatures would allow for deceleration of the erodant particles at a longer time and, therefore, the stresses generated will be smaller, which means less damage to the substrate. Second, the increase in temperature can affect the elastoplastic response of PU in such a way as to revert to its initial condition with less plastic deformation after the loading caused by the impact force. Thus, a higher number of impacts will be required to deform the PU to the detachment threshold, which means improved resistance to erosive wear. It was further shown that the increase in temperature can negatively affect the wear resistance in conditions where the final strength of PU becomes smaller than the stresses produced by the impact of erodant particles [21].

3. Modelling of wear process of polyurethane liners

3.1. Objectives in modelling of the wear process

Wear caused by impact and sliding of solid particles is a complicated process due to the high number of factors that affect the wear mechanism and final wear rate. Wear resistance is a function of (a) the properties of the erodant particles: shape, density, size, and hardness, (b) the properties of the target material: Young's modulus, plastic behavior, and final strength, and (c) wear testing parameters: velocity of impacting particle, angle of impacts, testing temperature, flow rate, sliding velocity, and pushing force. Accordingly, the experimental investigation of the effect of different parameters on the wear resistance is time consuming and costly. Moreover, wear caused by impact and sliding of erodant particles occurs within microseconds, and hence, the experimental study of the physics underlying the problem is not a trivial task [27]. The simulation of the wear process by analytical and numerical approaches not only enables the understanding of the fundamental principles of the wear mechanisms; the models developed can be employed after verification as predictive tools to study the effect of different parameters on the wear rate. Through different analytical and numerical models developed to date to simulate the wear of metals, ceramics, and elastomers, the finite element (FE) formulation has received great attention due to the potent formulation of this technique that enables the modelling of complex geometries, material models, and contact algorithms [27, 49, 50]. In the following section, the research works conducted for FE modelling of wear of PU elastomers will be reviewed and discussed.

3.2. Finite element modelling of the wear process

Several studies have focused so far on developing FE models for simulating the erosion caused by solid particle impact of ductile metals such as AISI 4140 steel and nickel (Ni), Al6061-T6, Ti-6Al-4V and brittle ceramics such as tungsten carbide (WC), Cr3C2, and SiC [50–53]. The models developed enabled an in-depth study of the stresses and strains produced during the erosion process and also assess the effect of testing factors such as particle size, shape, velocity, and impact angle on the erosion rate. On the other hand, fewer studies have focused on FE modelling of the solid particle erosion of soft elastomeric materials such as PU. In a recent study, Zhang et al. [43] simulated the impact of a single particle on PU liners by a FE modelling approach. In this model, an isotropic hardening elastic-plastic constitutive law was selected for the material formulation, and material removal was modelled by deleting the elements that exceeded the failure strain of the PU. The defined element removal criterion enabled to calculate the wear rate as a result of impact of a single erodant particle. The model developed was employed to study the effect of liner thickness on the erosion resistance of PU. The results obtained by the FE model were in good agreement with experiments only up to the liner thickness in which the effect of temperature was negligible. The model failed to correctly predict the same trend as experiments for the erosion rate versus liner thickness due to the fact that the model did incorporate the temperature rise caused by the repeated impact of particles.

The high elongation at break of PU elastomers can lead to significant deformation and distortion of elements in FE modelling of the wear process. To that end, FE-mesh–free techniques may be employed to eliminate the adverse effects of element distortion while modelling the erosive wear of soft substrates such as elastomers [54]. In mesh-free techniques, there is no connection between the nodes, and the model is discretized with scattered particles. For example, Gong et al. [54] developed a 3D combined FE-mesh–free model with smoothed hydrodynamics (SPH) particles. The viscoplastic material model of Johnson-Cook formulation was selected as for the material model. Even though the elements experienced extensive deformation in the FE model, the results obtained by the combined FE-SPH model predicted similar equivalent stresses at the impact point to that of the FE model with negligible difference [54]. Since the computation time by the FE model was approximately four times shorter than the combined the FE-SPH model, the FE formulation can be considered as the superior technique.

In a recent study by Ashrafizadeh [49], a more comprehensive material formulation capable of accounting for hyperelastic, elastoplastic, and stress softening of the elastomer was developed using the FE technique. The material model formulation successfully predicted the elastoplastic and stress softening response of the PU as validated by conducting cyclic deformations of a single element and comparing the stress-strain behavior of the element with that of experiments. This research allowed for an in-depth evaluation of the effect of temperature, material softness, final strength, and elastoplastic behavior on the stresses produced as a result of the impact of erodant particles. The model successfully simulated the cutting mechanism caused by the impact of a single erodant particle (see **Figure 9**). Moreover, the impact of ten solid particles was modelled to study the mechanism of material removal by accumulation of residual strains up to the detachment of material. The model provided support for this mechanism and successfully predicted the shape of the formed asperities (see **Figure 10**) similar to those of the eroded surface of PU as was observed from experiments (see **Figure 8**).

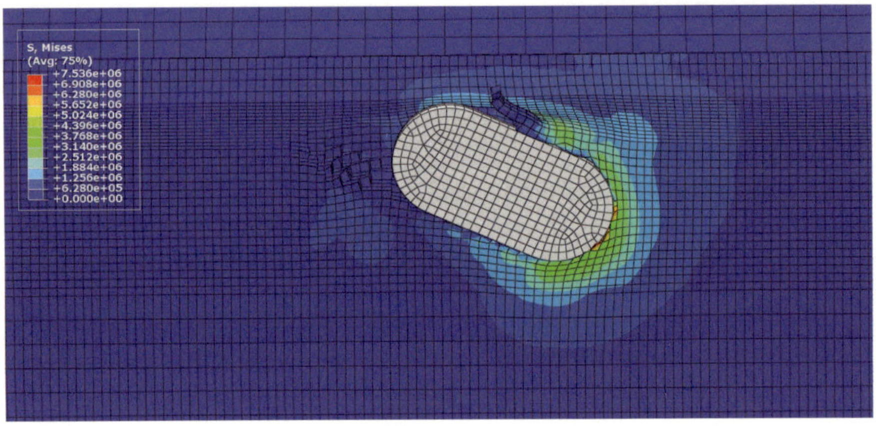

Figure 9. Material removal by cutting mechanism as predicted by the FE model [49].

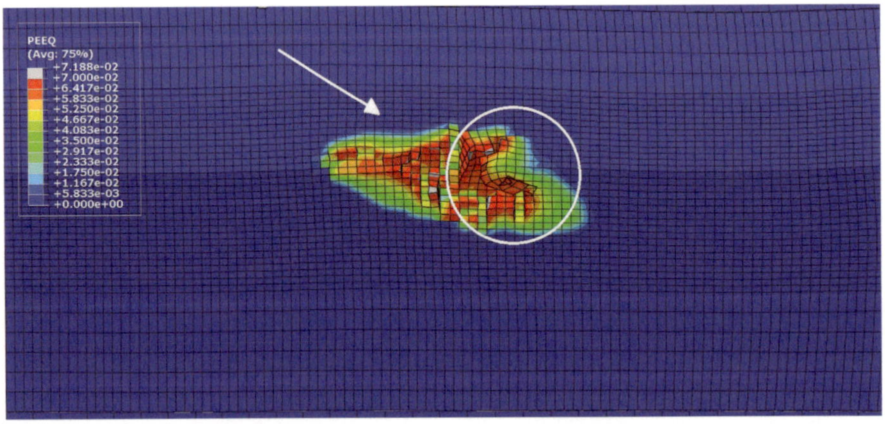

Figure 10. The ridges formed on the surface of an eroded PU as predicted by FE model [49].

4. Future trends

4.1. Wear resistance of filler modified polyurethane liners

Most of polymers including PU elastomers have low thermal conductivity of usually below 0.5 W/mK [31, 55]. Although the low thermal conductivity has made the PU an excellent option for insulation purposes, it can limit the applications of PU as protective wear resistant liners. As was discussed in Section 2.4 of chapter, heat may be produced during wear of PU elastomers by friction forces and hysteresis. The low thermal conductivity of the PU can allow for a localized temperature rise below the worn surface and, therefore, decrease in strength of the material leading to accelerated wear of the surface. On the other hand, fillers can be added to the structure of the polymer to develop multifunctional filler modified polymers with improved mechanical and thermal properties [31, 56].

It has been shown in previous studies that the thermal conductivity of polymers improved by means of addition of highly conductive powder materials such as metallic, ceramic, and carbon-based fillers with high thermal conductivity [57]. The improvement in thermal conductivity of the filler modified polymer is a function of the filler type, shape, size, and distribution within the polymer matrix [31, 58]. In a recent study by Akram et al. [31], nano-graphene fillers were added into an abrasion-resistant PU liner to improve its thermal conductivity. It was found that the addition of 4 wt.% of nanographene filler to the PU structure increased the thermal conductivity of the PU by 164%. Although the possible improvement in thermal conductivity of PU protective liners has been shown in previous studies, the effect of the added fillers on the abrasive and erosive wear resistance of elastomers has been the subject of fewer studies, and future work is required. An investigation on size, shape, concentration, and type of fillers that can simultaneously improve the thermal and mechanical properties of PU elastomers is suggested for future work.

4.2. Relation between wear resistance and viscoelastic response of polyurethane elastomers

In most of the previous studies, the wear resistance of PU elastomers has been compared with the mechanical properties that have been obtained in quasi-static or low strain rate testing conditions. However, the impact and rebounding of the particle occur within a few microseconds with PU deformation at very high strain rates. The mechanical response of PU elastomers at high strain rates can be significantly different from that of loadings at low strain rates [59]. PU may become harder at higher strain rate loading [59]. Thus, knowledge about the mechanical response of PU at high strain rates, and, in general, the viscoelastic response of PU elastomers can provide a deeper understanding of how the time-dependent properties of PU elastomers may affect the mechanism of material removal and the final erosion rate. The viscoelastic properties of PU elastomers can be characterized by hardness, DMA, creep, relaxation, and tensile/compression testing. A study of the relation between the viscoelastic properties and wear resistance of PU is suggested for future research.

4.3. Finite element modelling of the abrasive wear

Even though the successful FE modelling of the erosive wear has allowed for a better understanding of the stresses produced during the erosion process and material removal mechanism of PU elastomers, the number of studies that focused on FE modelling of the abrasive wear of elastomers is limited. Although in some studies such as the research by Martinez et al. [14], the FE technique was employed to analyze the abrasion phenomenon as a result of sliding and contact of PU over rough, rigid surfaces, the modelling of wear caused by abrasive, erodant particles is limited. Such models can allow for a better understanding of the mechanism of a material removal during abrasive wear of PU elastomers and identification of the key parameters affecting the abrasive resistance of PU. A research study to develop FE models to simulate the abrasion caused by sliding of solid particles is suggested for future work.

Author details

Hossein Ashrafizadeh, Andre McDonald and Pierre Mertiny*

*Address all correspondence to: pmertiny@ualberta.ca

Department of Mechanical Engineering, University of Alberta, Edmonton, Canada

References

[1] James B, Hudgins A. Failure analysis of oil and gas transmission pipelines. In: Makhlouf ASH, Aliofkhazraei M, editors. Handbook of Materials Failure Analysis with Case Studies from the Oil and Gas Industry. Waltham, MA, USA: Elsevier Ltd.; 2016

[2] Bariha N, Mishra IM, Srivastava VC. Hazard analysis of failure of natural gas and petroleum gas pipelines. Journal of Loss Prevention in the Process Industries. 2016;**40**:217-226

[3] Wang Z, Liu B, Yang Y, Han N, Zeng G, Ren A, Zhang T. Experimental and numerical studies on corrosion failure of a three-limb pipe in natural gas field. Engineering Failure Analysis. 2016;**62**:21-38

[4] Bai Y, Bai Q. Subsea Pipeline Integrity and Risk Management. Waltham, MA, USA: Elsevier Ltd.; 2014

[5] Al-Hamed A, Al-Fadhli HY, Al-Mutairi S, Yilbas BS, Hashmi MSJ, Stokes J. Investigation of HVOF thermal sprayed nanostructured WC-12Co mixed with Inconel-625 coatings for oil/gas applications. WIT Transactions on Engineering Sciences. 2013;**78**:215-225

[6] Ashrafizadeh H, Karimi M, Ashrafizadeh F. Failure analysis of a high pressure natural gas pipe under split tee by computer simulations and metallurgical assessment. Engineering Failure Analysis. 2013;**32**:188-201

[7] Bayram TC, Orbey N, Adhikari RY, Tuominen M. FP-based formulations as protective coatings in oil/gas pipelines. Progress in Organic Coatings. 2015;**88**:54-63

[8] Hu X, Barker R, Neville A, Gnanavelu A. Case study on erosion-corrosion degradation of pipework located on an offshore oil and gas facility. Wear. 2011;**271**:1295-1301

[9] Martínez D, Gonzalez R, Montemayor K, Juarez-Hernandez A, Fajardo G, Hernandez-Rodriguez MAL. Amine type inhibitor effect on corrosion-erosion wear in oil gas pipes. Wear. 2009;**267**:255-258

[10] Zavareh MA, Sarhan AADM, Zavareh PA, Basirun WJ. Electrochemical corrosion behavior of carbon steel pipes coated with a protective ceramic layer using plasma and HVOF thermal spray techniques for oil and gas. Ceramics International. 2016;**42**:3397-3406

[11] Li J, Hutchings IM. Resistance of castable polyurethane elastomers to solid particle erosion. Wear. 1990;**135**(2):293-303

[12] Arnold JC, Hutchings IM. The mechanism of erosion of unfilled elastomers by solid particle impact. Wear. 1990;**138**:33-46

[13] Clemitson IR. Castable Polyurethane Elastomers. Broken Sound Parkway, NW, USA: CRC Press, Taylor and Francis Group; 2008

[14] Martinez FJ, Canales M, Bielsa JM, Jimenez MA. Relationship between wear rate and mechanical fatigue in sliding TPU-metal contacts. Wear. 2010; **268**:388-398

[15] Budinski KG. Guide to Friction, Wear and Erosion Testing. West Conshohocken, PA, USA: ASTM International; 2007

[16] Zhang SW, Deguo W, Weihua Y. Investigation of abrasive erosion of polymers. Journal of Materials Science. 1995; **30**:4561-4566

[17] Ping Z, Youwei L, Chengqing Y, Jian L. Erosion behaviors of polymer coatings. In: Proceedings of CST2008 & ITS-IFToMM; Beijing, China; 2008. pp. 738-741

[18] Mertiny P, Popella R, Juss K. Manufacturing of large-scale polyurethane lined FRP piping. In: Proceedings of the ASME 2011 Pressure Vessels & Piping Division. Conference; Baltimore, Maryland, USA; 2011. Paper # PVP2011-57870

[19] Zafar S, Sharma AK. Abrasive and erosive wear behavior of nanometric WC-12Co microwave clads. Wear. 2016;**346-347**:29-45

[20] Stachowiak GW, Batchelor AW. Engineering Tribology. 3rd ed. Burlington, MA, USA: Elsevier Inc.; 2005

[21] Ashrafizadeh H, Mertiny P, McDonald A. Evaluation of the effect of temperature on mechanical properties and wear resistance of polyurethane elastomers. Wear. 2016;**368-369**:26-38

[22] ASTM Standard D2240, Standard Test Method for Rubber Property-Durometer Hardness. West Conshohocken, PA, USA: ASTM International; 2010

[23] Low IM. Effects of load and time on the hardness of a viscoelastic polymer. Materials Research Bulletin. 1998;**33**(12):1753-1758

[24] Suwanprateeb J. A comparison of different methods in determining load- and time-dependence of Vickers hardness in polymer. Polymer Testing. 1998;**17**:495-506

[25] ASTM Standard E384, Standard Test Method for Knoop and Vickers Hardness of Materials. West Conshohocken, USA: ASTM International; 2011

[26] ASTM Standard D638, Standard Test Method for Tensile Properties of Plastics. West Conshohocken, PA, USA: ASTM International; 2014

[27] ElTobgy MS, Ng E, Elbestawi MA. Finite element modeling of erosive wear. International Journal of Machine Tools & Manufacture. 2005;**45**:1337-1346

[28] Prisacariu C. Polyurethane Elastomers-From Morphology to Mechanical Aspects. Springer Science & Business Media; 2011

[29] ASTM Standard D2632, Standard Test Method for Rubber Property-Resilience by Vertical Rebound. West Conshohocken, PA, USA: ASTM International; 2017

[30] Menard KP. Dynamic Mechanical Analysis, A Practical Introduction. CRC Press; 1999

[31] Akram U, Mertiny P, McDonald A. Thermal characterization of NGP filler-modified polyurethane. In: Canadian International Conference on Composite Materials (CANCOM); Edmonton, AB, Canada; 2015

[32] ASTM Standard G75, Standard Test Method for Determination of Slurry Abrasivity (Miller Number) and Slurry Abrasion Response of Materials (SAR Number). West Conshohocken, USA: ASTM International; 2013

[33] ASTM Standard B611, Standard Test Method for Determining the High Stress Abrasion Resistance of Hard Materials. West Conshohocken, USA: ASTM International; 2005

[34] ASTM Standard G65, Standard Test Method for Measuring Abrasion Using the Dry Sand/Rubber Wheel Apparatus. West Conshohocken, USA: ASTM International; 2015

[35] ASTM Standard G76, Standard Test Method for Conducting Erosion Tests by Solid Particle Impingement Using Gas Jets. West Conshohocken, USA: ASTM International; 2013

[36] Zuev YS, Chelmodeev AD. Effect of temperature and concentration of aggressive media on rubber wear in an abrasive flow. Polymer Mechanics. 1968;**4**(1):70-74

[37] Marei AI, Izvozchikov PV. Determination of the wear of rubbers in a stream of abrasive particles. In: James DI, editor. Abrasion of Rubber. London, England: Maclaren; 1967. pp. 274-280

[38] Hill DJ, Kileen MI, O'Donnell JH, Pomery PJ, St John D, Whittaker AK. Laboratory wear testing of polyurethane elastomers. Wear. 1997; **208**:155-160

[39] Pitman JS. Proceedings of the 3rd International Conference on Internal and External protection of pipes. Cranfield, England: BHRA Fluid Engineering; 1979. p. 129

[40] Hutchings IM, Deuchar DWT. Erosion of unfilled elastomers by solid particle impact. Journal of Materials Science. 1987;**22**:4071-4076

[41] Beck RA, Truss RW. Effect of chemical structure on the wear behavior of polyurethane-urea elastomers. Wear. 1998;**218**(2):145-152

[42] Ashrafizadeh H, Mertiny P, McDonald A. Development of a test assembly for evaluating the erosion resistance of polyurethane elastomers at controlled temperatures. In: First Pacific Rim Thermal Engineering Conference (PRTEC); Hawaii's Big Island, HI, USA; 2016

[43] Zhang N, Yang F, Li L, Shen C, Castro J, Lee LJ. Thickness effect on particle erosion resistance of thermoplastic polyurethane coating on steel substrate. Wear. 2013;**303**(1-2):49-55

[44] Barkoula NM, Karger-Kocsis J. Review processes and influencing parameters of the solid particle erosion of polymers and their composites. Journal of Materials Science. 2002;**37**(18):3807-3820. [T84]

[45] Martinez FJ, Canales M, Izquierdo S, Jimenez MA, Martinez MA. Finite element implementation and validation of wear modeling in sliding polymer-metal contacts. Wear. 2012;**284-285**:52-64

[46] Zhang SW, He R, Wang D, Fan Q. Abrasive erosion of polyurethane. Journal of Materials Science. 2001;**36**(20):5037-5043

[47] Siegmann K, Mosch A, Carlesso A, Winkler M. Abrasive wear of selected plastics in solvent/corundum suspensions. Wear. 2015;**324-325**:118-121

[48] Jia X, Ling R. Two-body free-abrasive wear of polyethylene, nylon1010, expoxy and polyurethane coatings. Tribology International. 2007;**40**:1276-1283

[49] Ashrafizadeh H. Evaluation and analysis of the erosion performance of flame spray-coated polyurethane liners [thesis]. Edmonton, AB, Canada: University of Alberta; 2016

[50] Balu P, Kong F, Hamid S, Kovacevic R. Finite element modeling of solid particle erosion in AISI 4140 steel and nickel-tungsten carbide composite material produced by the laser-based powder deposition process. Tribology International. 2013; **62**:18-28

[51] Liu ZG, Wan S, Nguyen VB, Zhang YW. A numerical study on the effect of particle shape on the erosion of ductile materials. Wear. 2014;**313**:135-142

[52] Zhang H, Dong X. Finite element analysis of multiple solid particles erosion in cermet coating. Surface and Coatings Technology. 2015;**262**:184-190

[53] Wang YF, Yang ZG. Finite element model of erosive wear on ductile and brittle materials. Wear. 2008;**265**:871-878

[54] Gong Y, Yang ZG, Wang YF. Impact simulation on ductile metal pipe with polymer coating by a coupled finite element and meshfree method. Journal of Failure Analysis and Prevention. 2012;**12**:267-272

[55] Martienssen W, Warlimont H. Springer Handbook of Condensed Matter and Materials Data. Heidelberger, Berlin, Germany: Springer; 2005

[56] Wang X, Xing W, Song L, Yang H, Yeoh G. Fabrication and characterization of graphene-reinforced waterborne polyurethane nanocomposite coatings by the sol–gel method. Surface and Coatings Technology. 2012;**206**(23):4778-4784

[57] Fu YX, He ZX, Mo DC, Lu SS. Thermal conductivity enhancement with different fillers for epoxy resin adhesives. Applied Thermal Engineering. 2014;**66**:493-498

[58] Yu S, Yang S and Cho M. Analysis of thermal conductivity of polymeric nanocomposites under mechanical loading. Journal of Applied Physics. 2013;**114**:213503 1-9

[59] Sarva SS, Deschanel S, Boyce MC, Chen W. Stress-strain behavior of a polyurea and a polyurethane from low to high strain rates. Polymer. 2007;**48**:2208-2213

Chapter 7

Bio-Based Polyurethanes from Carbohydrate Monomers

Juan A. Galbis, María de Gracia García-Martín,
María Violante de Paz and Elsa Galbis

Additional information is available at the end of the chapter

http://dx.doi.org/10.5772/intechopen.69606

Abstract

The production of sustainable and environmentally friendly materials constitutes a growing field of attention. The incorporation of sugar-derived units into traditional step-growth polymers such as polyamides, polyesters, and polyurethanes is a healthy method to prepare novel biodegradable and biocompatible materials for application in the biomedical field and other sectors such as foodstuff packaging.

Keywords: monosaccharides, sugar-based monomers, synthesis, chemical structure, degradability, thermal properties

1. Introduction

The exhaustible nature of the oil reserves and the pollution that oil-based technological polymers cause on the environment because of their low degradability have intensified the interest in natural renewing resources for the chemical synthesis of polymers. Thus, sustained efforts have been extensively devoted to render new polymers based on natural renewing resources with higher degradability [1, 2].

Among the diverse natural sources, carbohydrates constitute highly convenient raw materials because they are inexpensive, readily available, and provide great stereochemical diversity. They are produced in large amounts by plants and microorganisms every year, and in some cases they even come from agricultural wastes. So, in the last few decades many research groups have been investigating on the preparation of new polymers, analogous to the more significant technical polymers, but based on monomers synthesized from natural and available sugars [3]. However, although several polymers have been synthesized using sugar-derived monomers with free hydroxyl groups [4–6], most syntheses of high-molecular-weight linear polymers involve derivatives having the hydroxyl groups appropriately blocked [3, 7].

Synthetic polymers obtained from sugar-based monomers are innocuous for human health. Their hydrophilic nature ensures a greater hydrolytic degradability [3], and reduces their environmental impact compared to classic polymers. Thus, the incorporation of sugar-derived units into traditional step-growth polymers constitutes an excellent approach to prepare novel biodegradable and biocompatible materials for application in the biomedical field and other sectors such as foodstuff packaging.

This chapter describes polyurethanes (PU) having the sugar units incorporated into the main chain. This topic has been partially reviewed before, but during the recent years numerous papers on the subject have been published. Thus, the following sections report on the syntheses, main properties, and applications of this type of sugar-based polymers that have been published mainly during the past decade. Patents have not been included so as to make the reference list more concise.

2. Synthesis of linear sugar-based polyurethanes (PU)

2.1. From diol-diisocyanate

Among the different procedures followed for the synthesis of [AABB]-type polyurethanes (PU), the most widely used is the classical isocyanate-based route, a polyaddition reaction of diols (or polyols) and diisocyanates (or polyisocyanates) (**Figure 1**, **Table 1**). This is a simple synthetic method conducted under inert atmosphere and in the absence of moisture at room temperature or above (from 25 to 80°C), either in solution or in bulk, with excellent conversions within 3–24 h for an extensive group of starting monomers. When the preparation of sugar-based PU is attended, the isocyanate-based method has been the route of choice so far, the sugar moiety being generally incorporated into the diol monomer.

Focusing on the preparation of sugar-based PU by the classical method, the solvents most commonly used are *N,N*-dimethylformamide (DMF) [11], *N,N*-dimethylacetamide (DMAc) [13], tetrahydrofurane (THF) [19] and, to a lesser extent, butanone [26], dimethylsulfoxide (DMSO), and hexamethylphosphoramide (HMPA) [31]. The polymerization can be catalyzed by a metal catalyst, the organotin catalyst dibutyltin dilaurate (DBTDL, **Figure 2**) [32] being the most widely used. Other metal catalysts are the commercial tin mercaptide esters registered as Metatin™ [18] and tetrabutyl titanate (TBT) [6]. Tertiary amines can also accelerate the polymerization process, and among them, the most extensively selected catalysts are 1,4-diazabicyclo[2.2.2]octane (DABCO) [10] and *N,N*-dimethylcyclohexylamine (DMCHA) [33]. Triethylamine (TEA) was chosen as (co-)catalyst in the preparation of segmented [27] and cross-linked [34] PU.

High temperatures and the incorporation of a liquid diol or polyol in the formulation, namely polyethylene glycol (PEG) [35–37], polyglycerol (PGL) [38], polypropylene glycol (PPG) [38, 39], and polycaprolactone (PCL) [40] are required when the polymerization is performed in bulk.

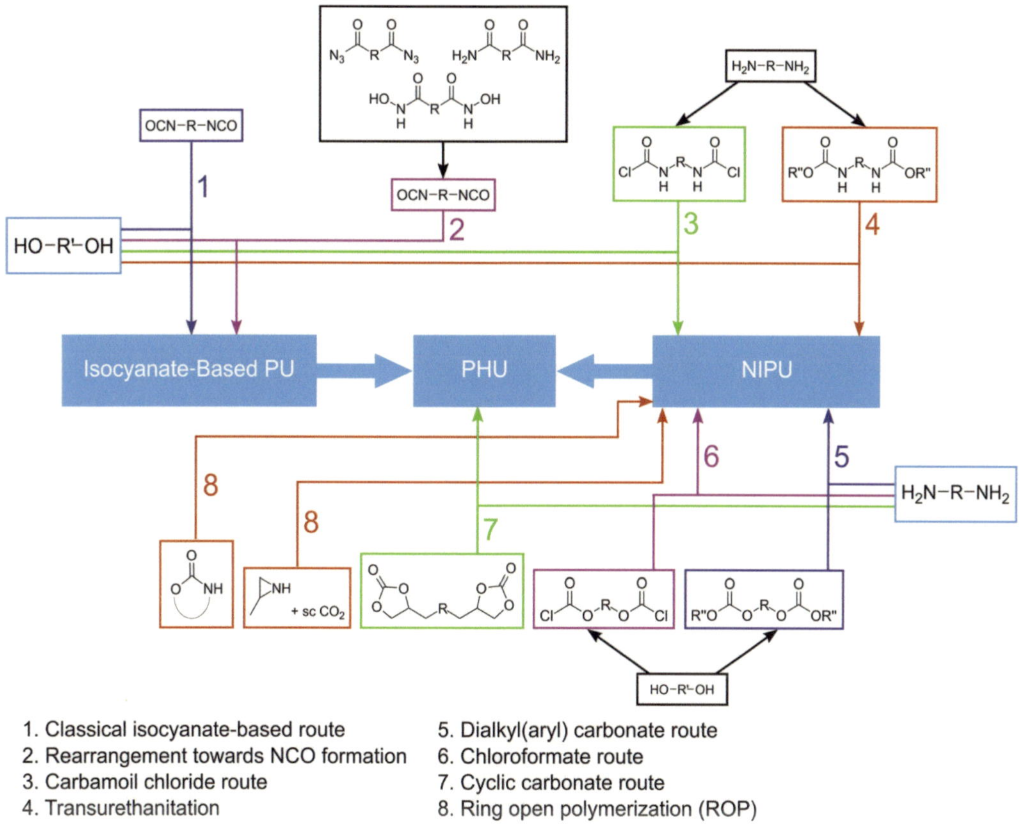

Figure 1. Most common synthetic routes to polyurethanes.

In sugar-based PU, the polyisocyanates and polyisothiocyanates most widely used are displayed in **Figure 3**, most of them being commercially available.

Other isocyanate-based routes are those in which a rearrangement of several acyl derivatives such as acyl azides (Curtius rearrangement), carboxamides (Hofmann rearrangement), and hydroxamic acid (Lossen rearrangement) conducts to the isocyanate monomer *in situ* (**Figure 1**). Thus, an article on the preparation of bio-renewable polyurethanes through Curtius rearrangement has been published. Initially, the synthesis of two new non-hindered diisocyanates based on isosorbide (**1**) and isomannide (**20**) without using petroleum-based reagents was attempted. The diisocyanates formation was carried out as follows: first, diols **1** or **20** were esterified with succinic anhydride in bulk at 120°C; second, the acidic derivatives were transformed into the corresponding diacid chlorides at low temperature, and lastly the diacid chlorides led to the diisocyanate derivatives **47** and **48** *via* a two-step Curtius rearrangement with overall conversions ranging from 52 to 60% [28]. Both diisocyanates were proved to be useful as starting materials in the preparation of two PU: a stereoregular PU (with D-*manno* configuration in all the monomers) and a non-stereoregular PU (with D-*gluco* configuration).

Diol/diamine	Diiso(thio)cyanates	Reaction conditions	T_g	Others	References
1 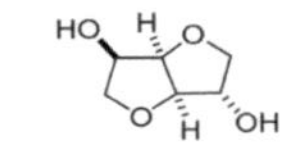	3 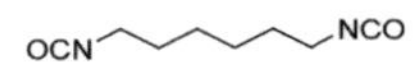	DMF, 25°C, 72 h, DBTDL	–		[8]
2 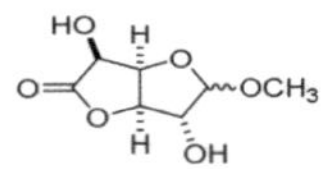	4 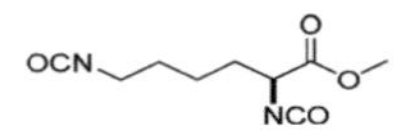				
5 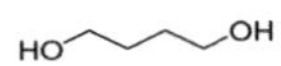	10 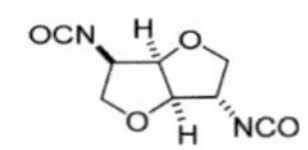	DMAc, 80°C, 24 h, DBTDL	From 48 to 112°C		[9]
6 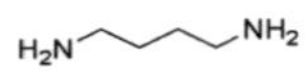	11 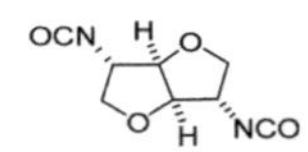				
7 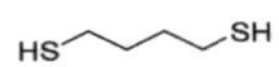	12 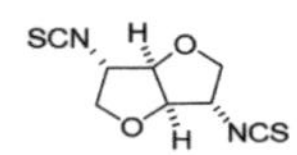				
8 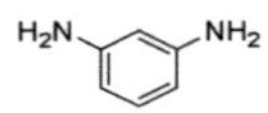	13 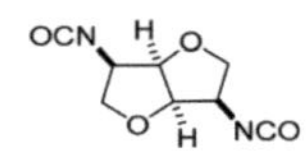				
9 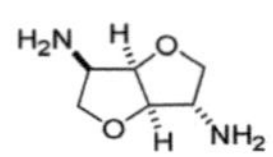					
1 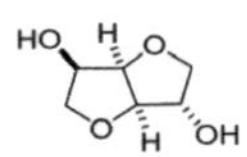	11 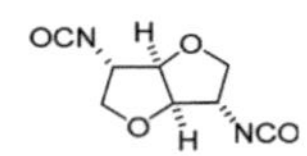	DMAc, 80°C, 24 h, DBTDL			[9]
	12 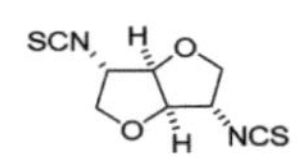				
14 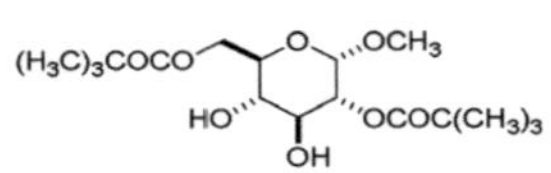	3 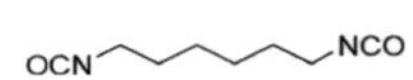	THF, reflux, 60 h, DABCO	–		[10]
15 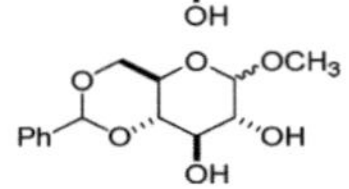					

Diol/diamine	Diiso(thio)cyanates	Reaction conditions	T_g	Others	References
16 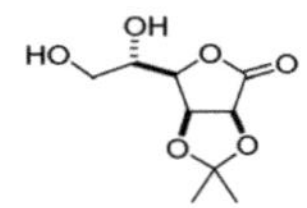	**3** 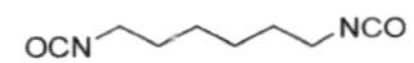	DMF, 25°C, 48 h, DBTDL	–	Desprotection → PHU	[11]
	4 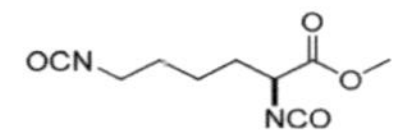				
17 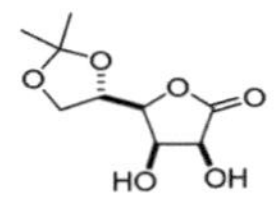					
18 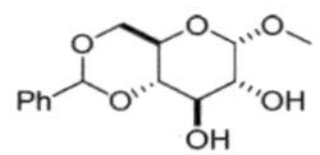	**3** 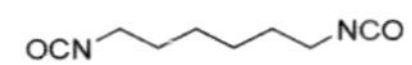	DMF, 25/50°C, 24/48/72 h, DBTDL		Desprotection → PHU	[12]
	4 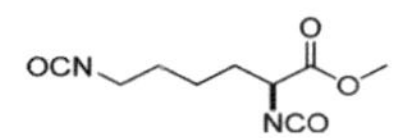				
19 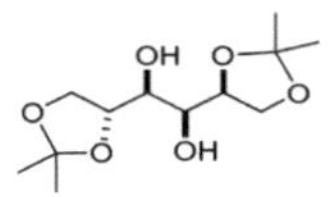					
1 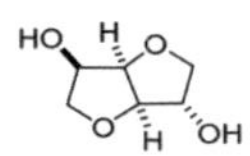	**3** 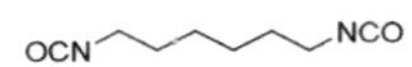	DMAc, 80°C, 24 h, DBTDL	From 77 to 183°C		[13]
20 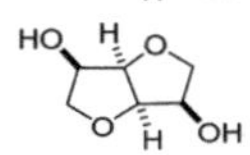	**23** 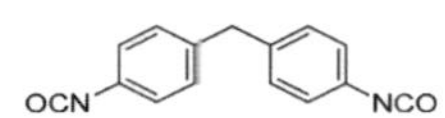				
21 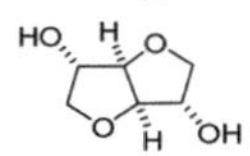					
22 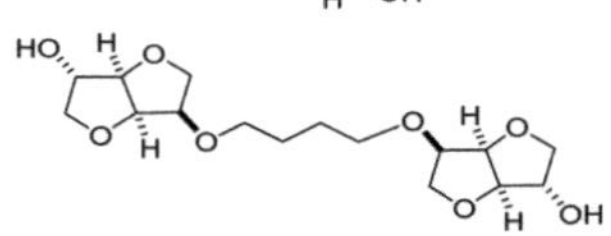					

Diol/diamine		Diiso(thio)cyanates		Reaction conditions	T_g	Others	References
24	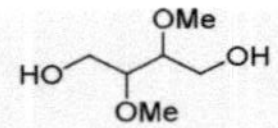	3	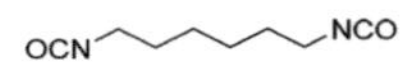	THF/DMF, 25/40°C, 3/24 h, DBTDL	From 22 to 110°C		[14]
25	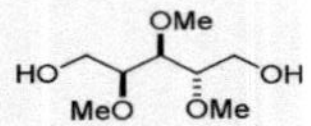	23	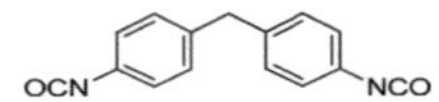				
26	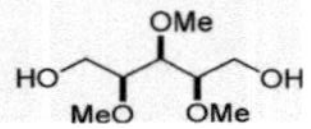						
27	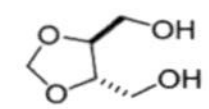	3		DMF, 40°C, 24 h, DBTDL Codiol: BD	From 35 to 126°C	Desprotection → PHU	[15]
28	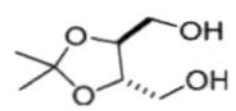	23	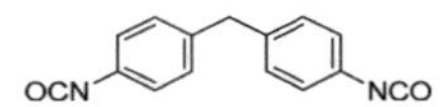				
29	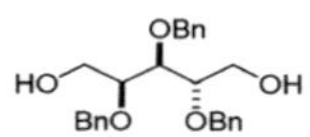	3	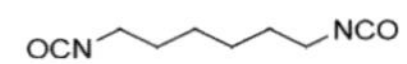	THF/DMAc, 25°C, 1/3 h, DBTDL	36°C, 108°C		[16]
30	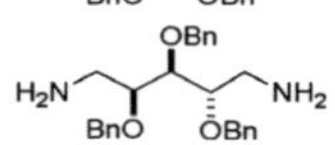						
31	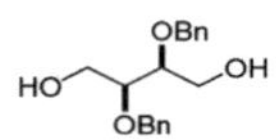	3	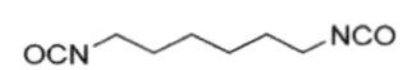	THF/DMF, 25°C, 3/24 h, DBTDL Codiol: BD, PD	From 21 to 79°C	Desprotection → PHU	[17]
29	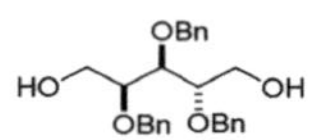	23	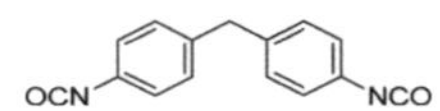				
32	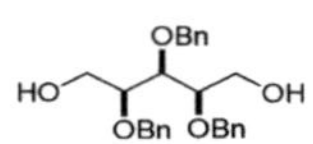						
1	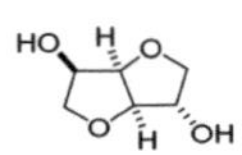	3	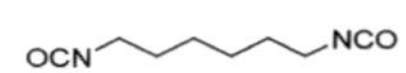	DMF/bulk, 25/130°C, 24 h, DBTDL/METATIN Codiol: PCL-3000/BD	From −60 to −3°C	Segmented co-PU	[18]
33	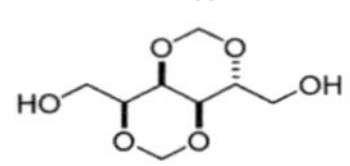	23	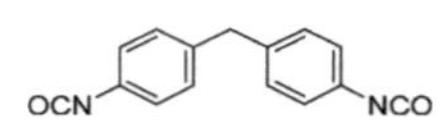				

Diol/diamine		Diiso(thio)cyanates		Reaction conditions	T_g	Others	References
34	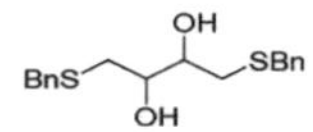	**3**	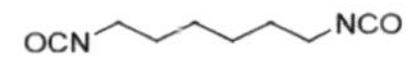	DMF/THF, 25/70°C, 3/5/7 h, DBTDL Codiol: TEG	From −15 to 23°C		[19, 20]
25	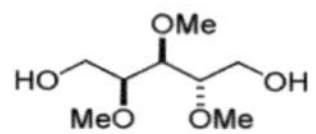	**3**	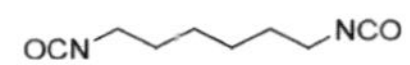	THF/DMAc, 25/−17°C, 1.5/3 h, DBTDL Codiol: DiT	From 1 to 36°C	.	[21]
29	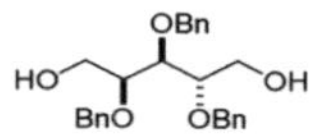						
35	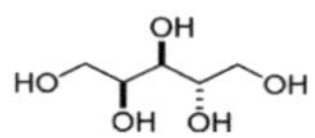						
36	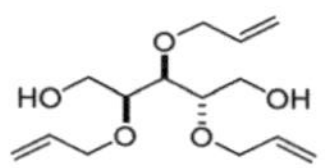	**3**	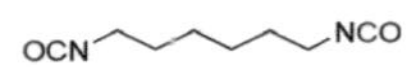	DMF, 60°C, 5 h, DBTDL	9°C, 16°C	Functionalization by thiol-ene click reaction → PHU	[22]
37	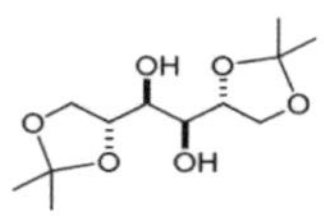	**3** **4**	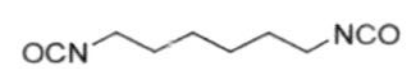 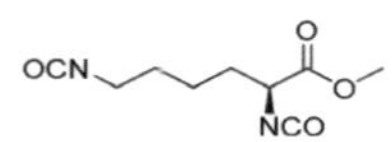	DMF, 50/70/80°C, 12/24 h, DBTDL Codiol: PTMG-650	–	Desprotection → PHU	[23]
38	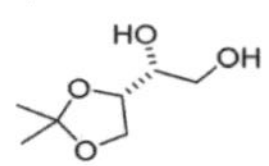						
39	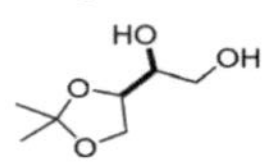						

Diol/diamine		Diiso(thio)cyanates		Reaction conditions	T_g	Others	References
36	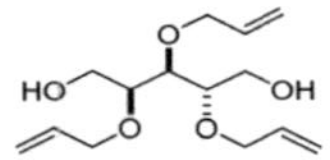	3	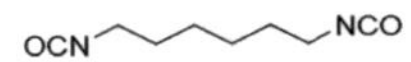	DMF, 25°C, 5 h, DBTDL Codiol: DiT	From −21 to 91°C	Functionalization by thiol-ene click reaction → PHU; → free COOH; → free NH_2	[24, 25]
40	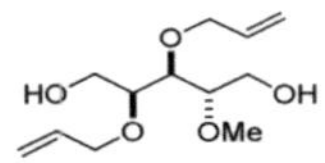	23	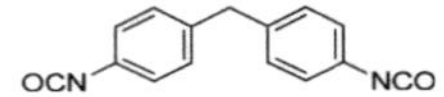				
41	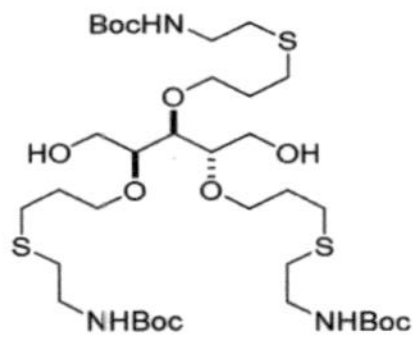						
42	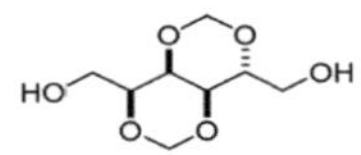	3	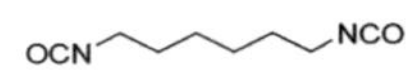	DMF, 40/60°C, 48 h, DBTDL	From 50 to 161°C	Desprotection → PHU	[7]
43	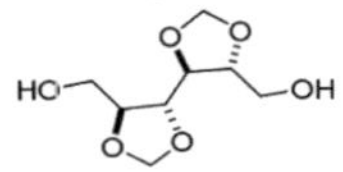	23	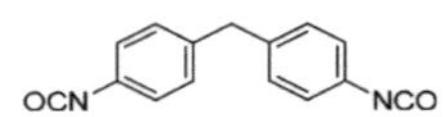				
44	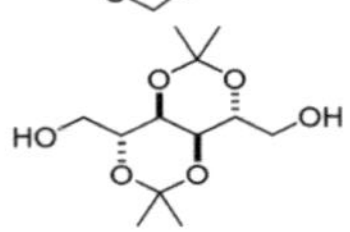						
1	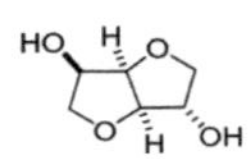	45	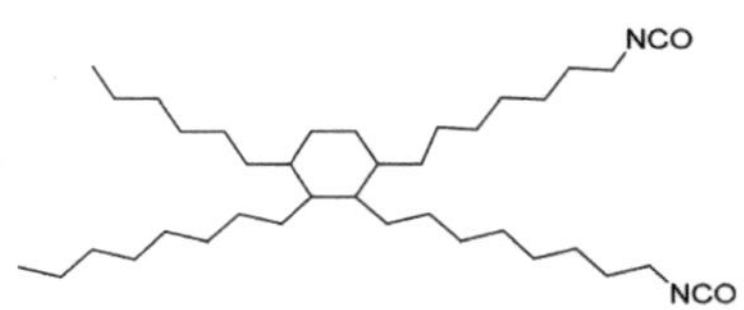	Butanone, 30/70°C, 5/7 h, DBTDL/TEA	From 18 to 58°C	Dispersion	[26, 27]
		46	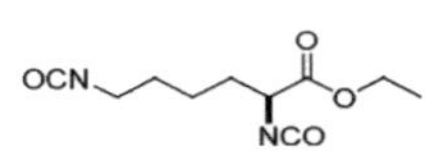				

Diol/diamine		Diiso(thio)cyanates		Reaction conditions	T_g	Others	References
1	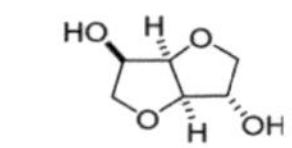	47	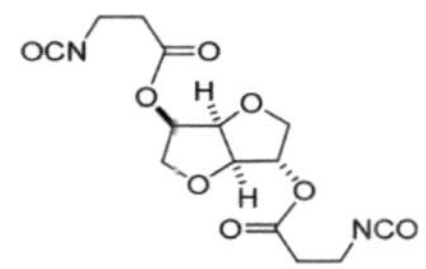	DMF, 120°C, 48 h, DBTDL	78°C, 81°C		[28, 29]
20	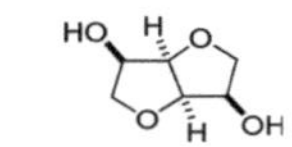	48	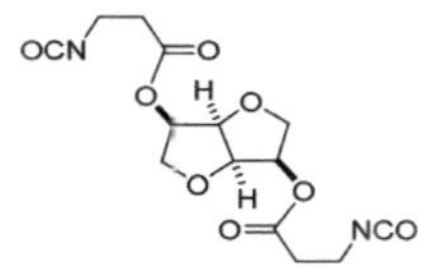				
49	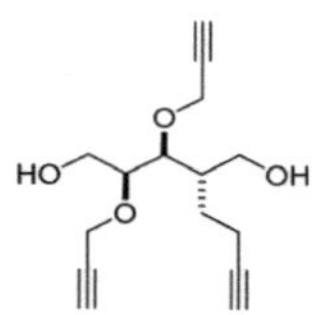	3	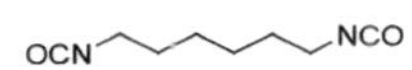	THF, 25°C, 24 h, DBTDL Codiol: DiT, BD, Octanediol	From −5 to 76°C	Alkyne-azide click reaction Dodecyl azide, Triethyleneglycol azide, PEG-750 azide	[30]

Table 1. Summarized literature about linear sugar-based PU prepared from diisocyanates.

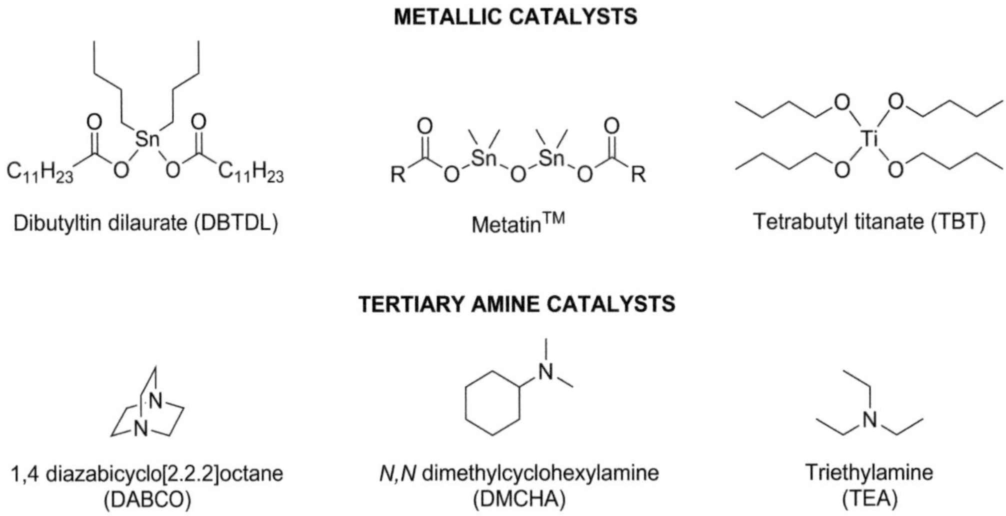

Figure 2. Catalysts for the synthesis of PU via diisocyanate-based polymerization.

To conduct the synthesis of linear [AB]-type homopolyurethanes, both the nucleophilic and the electrophilic groups (e.g., hydroxyl and isocyanate groups [32, 41, 42]) need to be present in one sole monomer (**Table 2**).

Thiem and coworkers were the first to attempt this approach with success in order to synthesize sugar-based PU. They published the preparation of 2-deoxy-1,4:3,6-dianhydro-2-isocyanato-L-iditol (**50**) from isosorbide (**Figure 4**), and the difunctional monomer proved to be suitable to polymerize in the presence of a catalyst [32].

The synthesis of a galactitol-based PU from the α,ω-hydroxyl isocyanate monomer **52** was carried out similarly to the abovementioned iditol-based PU. The preparation of the monomer 1-deoxy-1-isocyanato-2,3:4,5-di-*O*-isopropylidene-D-galactitol (**52**) was conducted from D-galactono-1,4-lactone by a four-step route. The monomer was then polymerized in THF in the presence of zirconium(IV) acetylacetonate [Zr(acac)$_4$] as catalyst. The removal of the acetal-protecting groups led to a galactitol-based polyhydroxyurethane (PHU) [41]. In addition, Kolender et al. synthesized a glucitol-based PU by polymerization of another α,ω-hydroxyl isocyanate monomer, the 1-deoxy-1-isocyanato-2,3,4,5-tri-*O*-methyl-D-glucitol (**53**) [42].

2.2. Eco-friendly methods (isocyanate or/and stannous free) to prepare PU

Although polymerization reactions of diisocyanates with diols are the main method to synthesize linear PUs, in the last few years and due to the toxicity of stannous catalysts and common aromatic diisocyanates such as 4,4'-methylenediphenyl diisocyanate (MDI) and toluene diisocyanate (TDI), a large number of works have been reported in literature on the synthesis of isocyanate-free PU, also called non-isocyanate PU (NIPU) (**Figure 1**, **Table 3**). Two interesting reviews have been recently published in which alternative pathways for PU syntheses are studied in detail [43, 44].

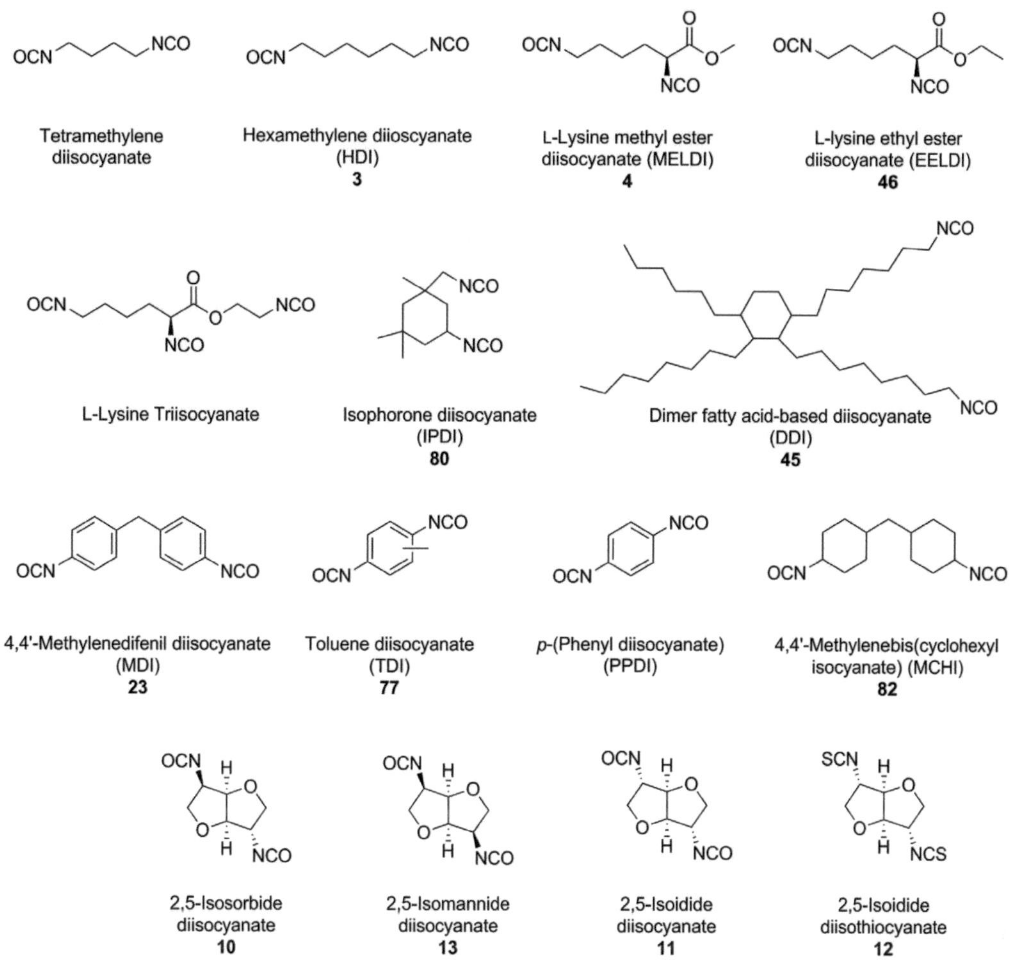

Figure 3. Chemical structure of the most common polyisocyanates and polyisothiocyanates used in the synthesis of sugar-based PU.

2.2.1. Dicarbamate and diols

One of the methods to use is the transurethanization polymerization between a dicarbamate and a diol in which the side product is an alcohol, usually of low-molecular weight. For example, linear polyurethanes with free hydroxyl groups have been successfully prepared by Galbis et al. [6] from xylitol (**55**) and the aliphatic carbamates dimethyl hexamethylenedicarbamate (HDC, **56**) or di-*tert*-butyl-4,4'-diphenyl methyl dicarbamate (MDC, **59**), to render NIPU with enhanced hydrophilic character.

2.2.2. Bis(chloroformate) and diamines

To avoid the use of diisocyanates, a variety of new materials can also be obtained by polycondensation of diamines with freshly prepared bis(chloroformate)s. For example, two reactive bis

Monomer	Solvent	T_g	Others	References
50 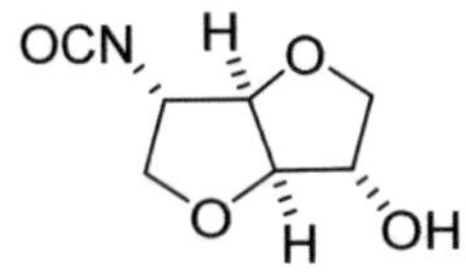	DMAc/MeOH, 80°C, DBTDL	–		[32]
51 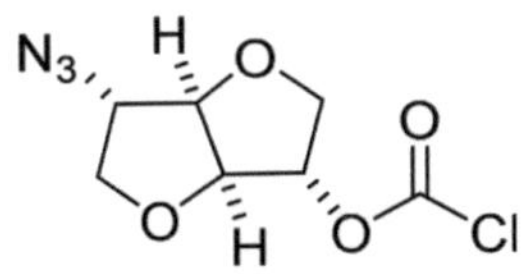				
52 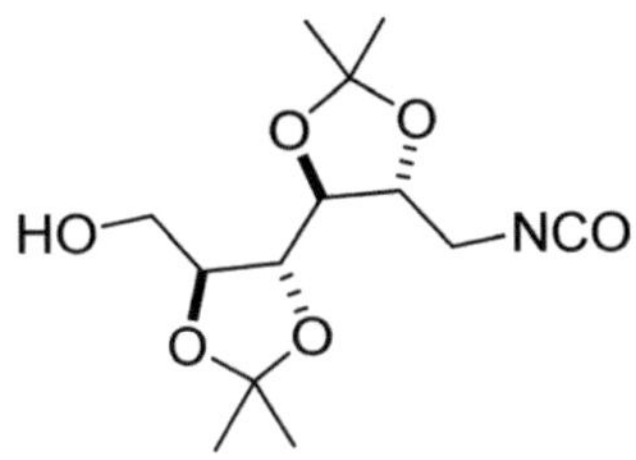	THF, 60°C, 48 h, Zr(acac)4	–	Desprotection → PHU	[41]
53 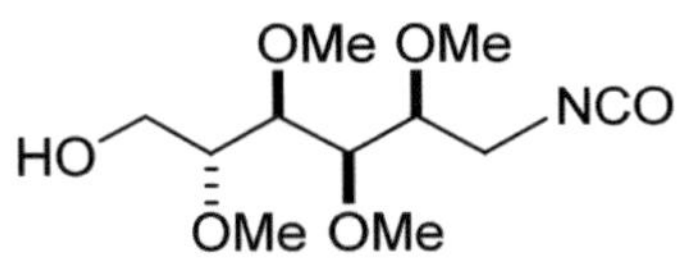	THF/DMF, 40°C, 16/72 h, Zr(acac)4	–		[42]
54 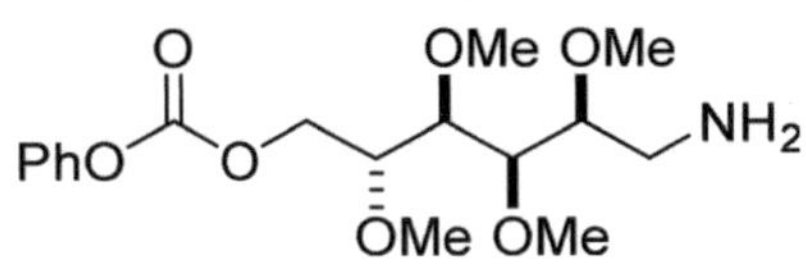				

Table 2. Selected examples of the use of a sole carbohydrate-based monomer for the preparation of PU.

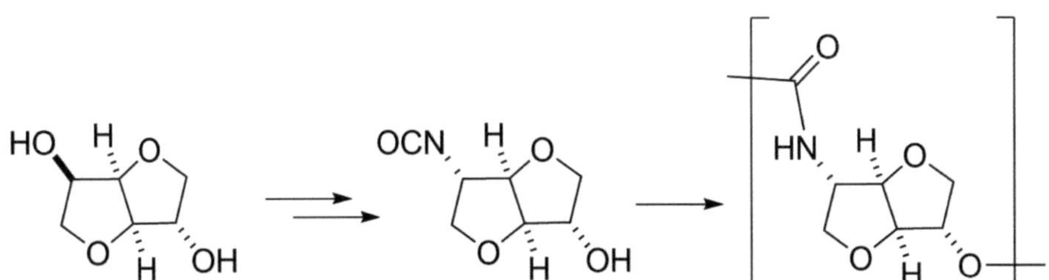

1 50

Figure 4. PU based on 2 deoxy-1,4:3,6-dianhydro-2-isocyanato-L-iditol (**50**).

(chloroformates) were formed from methyl ether diols of L-arabinitol and xylitol (**73, 74, Figure 5**). They were polymerized with commercial diamines (cystamine and cystine dimethyl ester) by interfacial polycondensation using sodium lauryl sulfate as surfactant [48]. The new

http://dx.doi.org/10.5772/intechopen.69606

Diol/diamine	Dielectrophile	Reaction conditions	T_g	Others	References
55 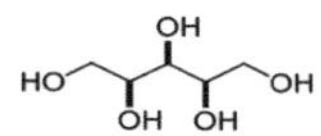	**3** 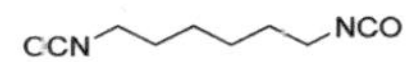	Bulk/DMF/DMSO/DMAc, −60/−20/25/100/170°C, 0.5/1/5/6 h, no cat/DBTDL/TBT	From −17 to 140°C	PHU	[6]
56 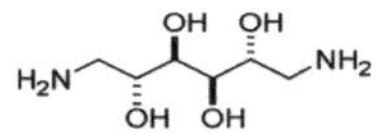	**23** 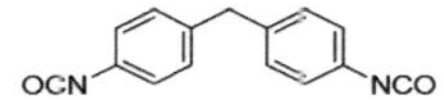				
57 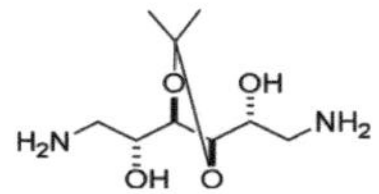	**58** 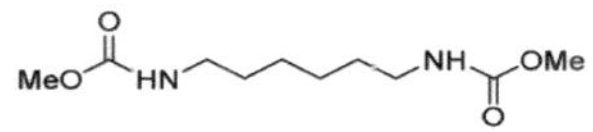				
	59 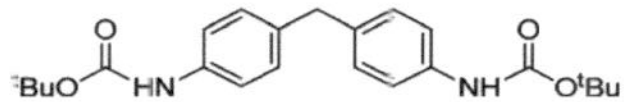				
60 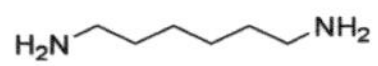	**61** 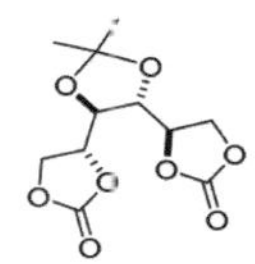	THF/Diglyme/DMAc, 25/45/65/80°C, 12/24 h, no cat.	From 64 to 79°C	PHU	[45]
	62 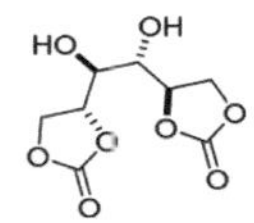				
63 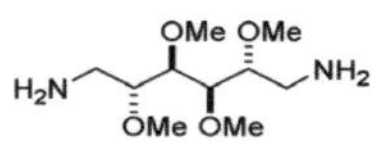	**65** 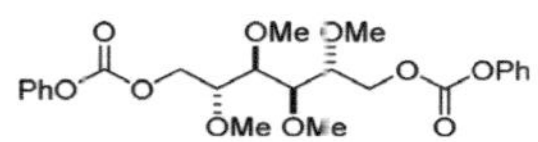	THF/DMF, 85°C, 2/4/7 h, no cat.	From 38 to 61°C		[46]
64 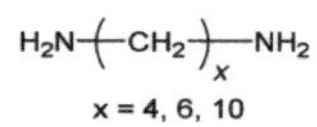					

Diol/diamine		Dielectrophile		Reaction conditions	T_g	Others	References
66	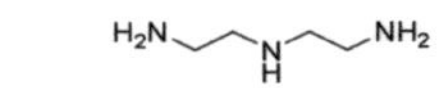	70	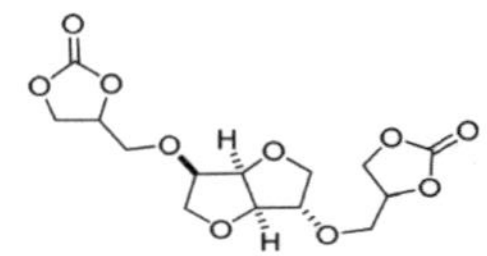	DMF, 25°C, 12 h, cat. (not disclosed)	From −8 to 59°C		[47]
67	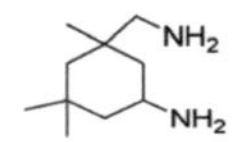						
68	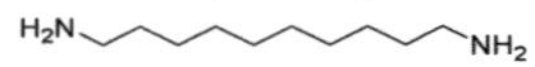						
69	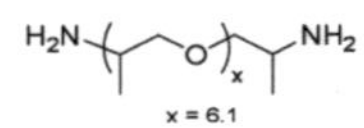						
71	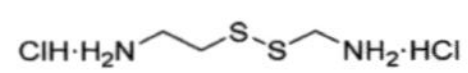	73	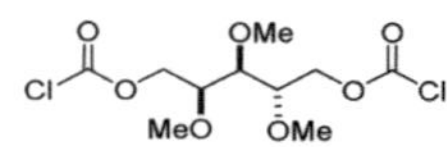	Toluene-water, 25°C, 0.5 h, sodium lauryl sulfate	From 35 to 63°C	Interfacial polycondensation	[48]
72	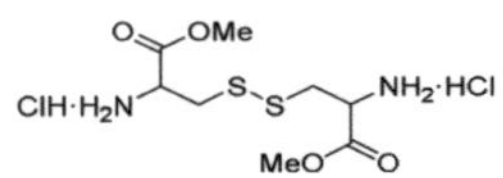	74	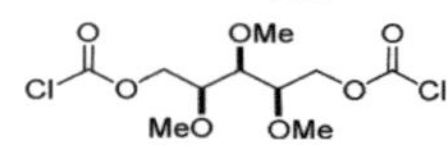				

Table 3. Summarized literature about linear sugar-based PU prepared by eco-friendly alternative pathways.

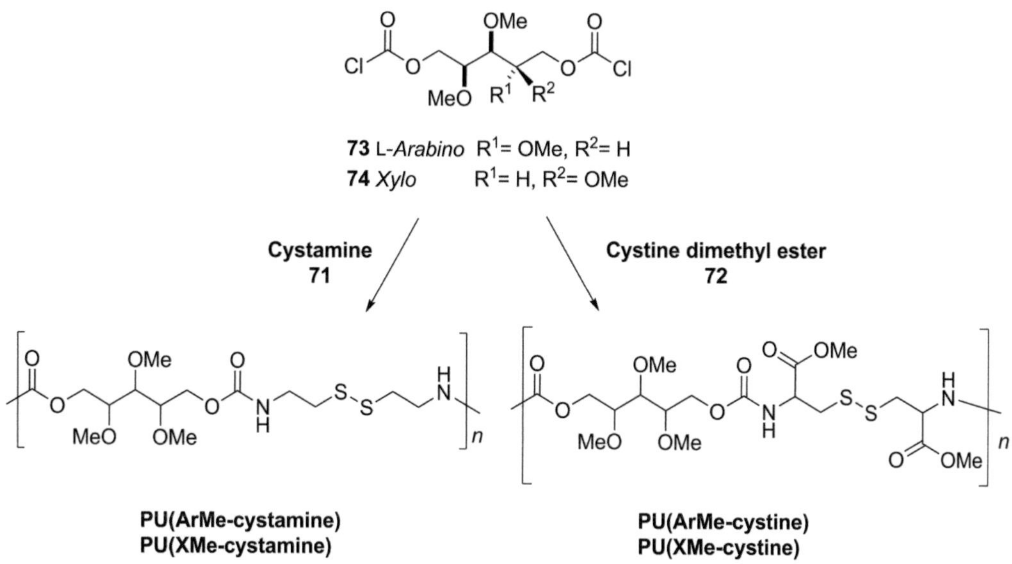

Figure 5. Reduction-sensitive homopolyurethanes from diamines and bis(chloroformate)s.

NIPU, bearing the labile disulfide bond in their structure, were degradable under reductive environments.

It was demonstrated that the presence of methoxycarbonyl side groups in the PU makes those materials more degradable, not only under hydrolytic conditions but also under reductive environments, probably due to their lower crystallinity and a better water/glutathione penetration in their structure. Moreover, differential scanning calorimetry (DSC) studies showed that the incorporation of the abovementioned pendant methoxycarbonyl groups into the homopolymers resulted in a more rigid material.

This synthetic method provides a useful tool toward the synthesis of chemical diverse NIPU because of the large set of commercial diamines available and the excellent on-hand chemical procedures for the synthesis of amines.

2.2.3. *Dialkyl(or aryl)carbonate and diamines*

The polycondensation reaction between dialkyl- and diarylcarbonates with diamines is another alternative method for the preparation of NIPU (**Figure 1**). Varela et al. reported the preparation of the diarylcarbonate 1,6-di-*O*-phenyloxycarbonyl-2,3,4,5-tetra-*O*-methyl-D-mannitol (**65**) and its use as comonomer in the preparation of some NIPU at 85°C [46].

The reaction of an α,ω-amino-arylcarbonate monomer, 1-amino-1-deoxy-2,3,4,5-tetra-*O*-methyl-6-*O*-phenyloxycarbonyl-D-glucitol hydrochloride (**54**), was tested for the preparation of sugar-based [AB]-polyurethane (**Figure 6**). Thus, the free hydroxyl group from the starting material was activated by means of the preparation of the phenylcarbonate derivative and further self-polymerized in THF in the presence of diisopropylethylamine (DIPEA), rending a D-glucitol-based PU with low-molecular weight [42].

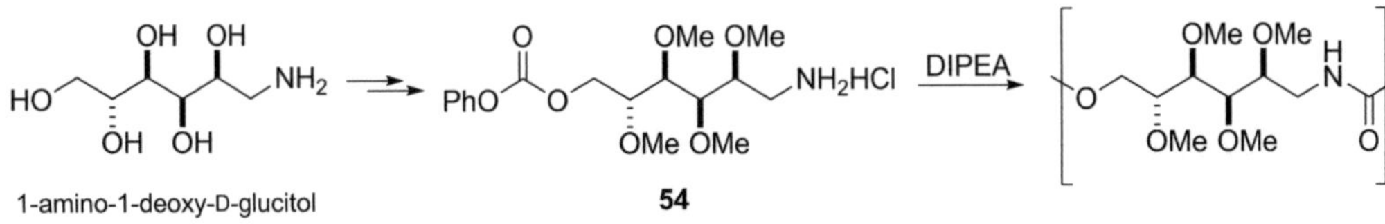

Figure 6. Synthesis of [AB]-polyurethane from 1-amino-1-deoxy-D-glucitol.

2.2.4. Dicyclocarbonates and diamines

Among the numerous pathways leading to NIPU, the polyaddition of cyclic carbonates with amines seems to be the most interesting route, and it is being widely investigated by numerous research groups (**Figure 1**). This method has been tested in sugar derivatives by Prömpers et al. They reported the preparation of PHU from D-mannitol-1,2:5,6-dicyclocarbonate **62** and its 3,4-*O*-isopropylidenene derivative **61** with hexamethylenediamine (**60**) [45].

Similarly, Besse et al. reported the preparation of linear and branched isosorbide-based polyhydroxyurethanes, with low T_g values (from −8 to 59°C) [47]. The isosorbide-based dicyclocarbonate monomer **70** (**Figure 7**) was prepared from a diepoxide according to the method previously described by Brocas et al. [49].

However, this route displays two major drawbacks: the low reactivity between cyclic carbonates and amines, and a limited degree of advancement of reaction during the room-temperature polymerization that leads to low-molecular-weight PHU. Consequently, highly polydisperse, low-molecular-weight materials were isolated.

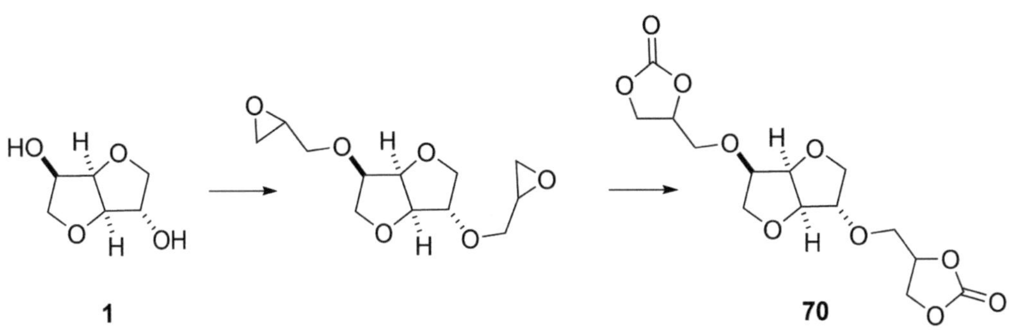

Figure 7. Isosorbide-based dicyclocarbonate monomer.

3. Synthesis of cross-linked sugar-based PU and segmented PU

For the preparation of cross-linked sugar-based PU, a monomer with functionality above two is required. Just one work was found in which the cross-linker was a triisocyanate derived from L-lysine derivative [35]. In general, the cross-linking is accomplished by the use of a mono- or a disaccharide with all its hydroxyl groups unprotected. Thus, glucose and sucrose are widely used in the formulations of carbohydrates-based networks [38–40, 50, 51] although

http://dx.doi.org/10.5772/intechopen.69606

Cross-linker	Dinucleophile	Dielectrophile	Reaction conditions	T_g	Others	References
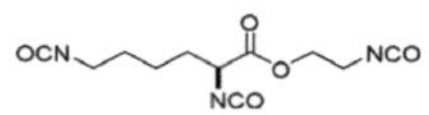	75 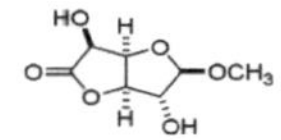	4 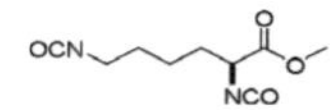	DMF, 27°C, 72 h, DBTDL Codiol: PEG-1000	–	Gel	[35]
glucose, sucrose 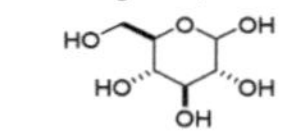 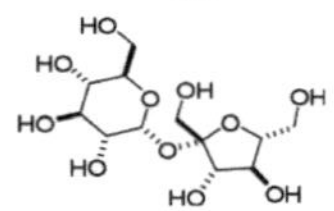	76 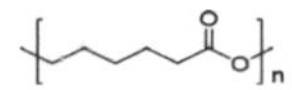	77 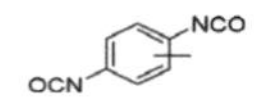	Bulk, 60/100°C, 1.5/48 h, no cat. Codiol: BD	–		[40, 50]
sucrose 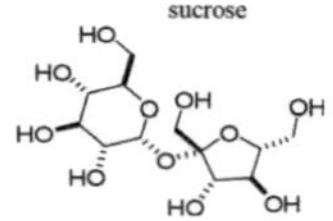	78 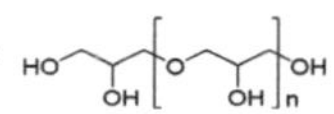	23 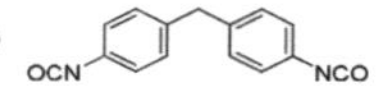	Bulk, –, –, DMCHA	–	Foam	[33]
glucose, sucrose, starch 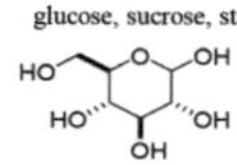 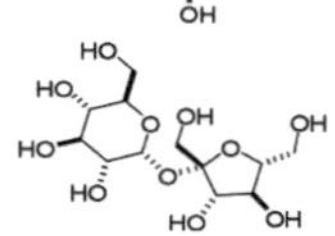	79 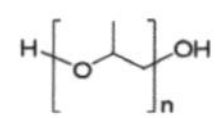	3 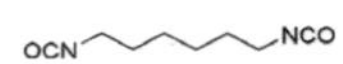	Bulk, 20/80°C, 3 h, DBTDL	–	Foam	[38]

Cross-linker	Dinucleophile	Dielectrophile	Reaction conditions	T_g	Others	References
sucrose dihidroxy oleate, methylglucopyranose dihidroxy oleate 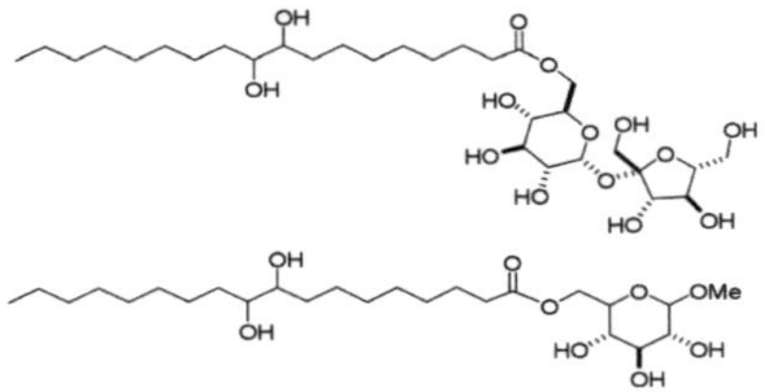		80 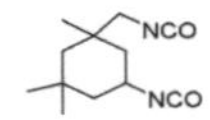	THF/DMF, 60°C, 2/20 h, DBTDL	39°C, 111° C	Linear or cross-linked depending on the solvent	[51]
sucrose, maltose 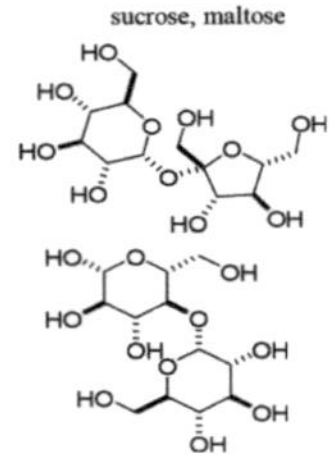	81 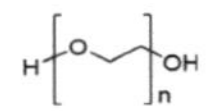	82 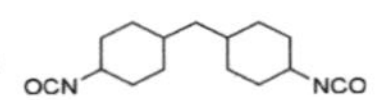	THF, 90°C, 5 h, no cat.	From −38 to −8°C	Adhesives	[36]
lignin, xylaric acid (and lactone) methyl esther 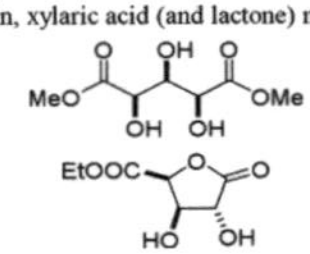	81 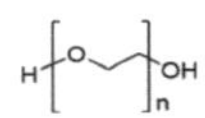	23 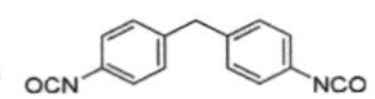	DMF, 20/40/60°C, 25 h, DBTDL Codiol: PEG-1500	From 95 to 132°C	Film	[37]

http://dx.doi.org/10.5772/intechopen.69606

Cross-linker	Dinucleophile	Dielectrophile	Reaction conditions	T_g	Others	References
glucose, sucrose, cellulose, starch, castor oil 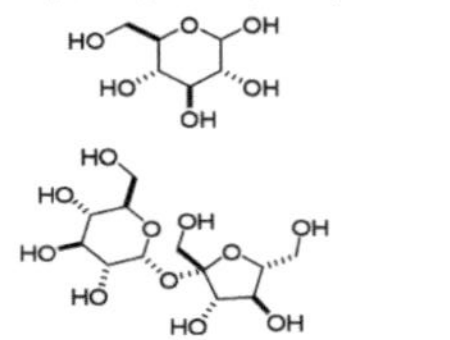	79 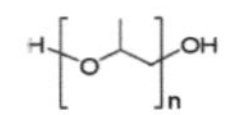	77 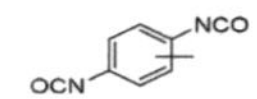	THF, 25/80°C, 6 h, DBTDL	From −58 to −31°C	Film	[39]
xylopyranose 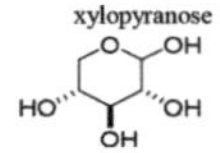	81 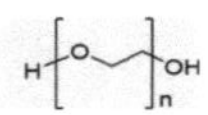	82 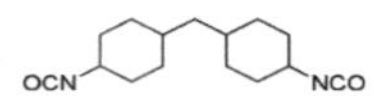	THF-DMF 9:1, 90°C, 8 h, TEA	From −58 to −18°C	Tissue adhesive	[34]

Table 4. Selected examples of the use of monosaccharides and disaccharides for the preparation of cross-linked PU.

other monosaccharide and disaccharide derivatives such as xylaric acid [37], maltose [36], and glucosides [51] have also been chosen (**Table 4**). For example, Ates et al. [34, 36] prepared some non-aromatic cross-linked polyurethanes with potential use as surgical tissue adhesives. PEG diol **81** and 4,4'-methylenebis(cyclohexyl isocyanate) (MCHI, **82**) were polymerized with a certain amount of a sugar (maltose, sucrose, or xylose) as cross-linker. The reaction was carried out in THF or DMF-THF mixtures, at high temperatures in the absence of catalyst or adding triethylamine (TEA). The material cross-linked with xylose 55 [34] displayed high adhesiveness and biocompatibility properties, making it suitable for being used in the medical field. Depending on the final use of the cross-linked material, the reaction can be conducted in bulk [33, 38, 40, 50].

The use of flexible aliphatic polyether/polyester chains such as PCL **76** [18], PGL **78** [33], PEG **81** [35–37], PPG **79** [38, 39], or oleic ester derivatives [51] (in combination with other diols) can lead to segmented PU as well as carbohydrates-based cross-linked materials with differentiated regions. The synthesis can be carried out "one pot," in which the polyols are incorporated into the polymer feed together with the sugar-based monomers and the other components. For example,the synthesis of highly functionalized low-molecular-weight polyether-polyols initiated by PGL and mixtures of PGL and sucrose has been described [33]. In the "one pot" preparation of rigid polyurethane foams in bulk and using MDI as diisocyanate, the polyether and/or polyols constituted the soft segment.

Even though materials obtained "one pot" displayed good physical and mechanical properties, excellent dimensional stability as well as low friability, an improvement of microphase separation can be accomplished when the polymerization is performed in two stages and the flexible polyol is incorporated in the first step [23, 26, 27]. The last step promotes the formation of urethane or urea linkages, in the so-called hard region with a high density of hydrogen bonds that provides stiffness to that section and elastomeric properties to the final product.

4. Synthesis of PU using bicyclic monomers derived from 1,4:3,6-dianhydrohexitols

The 1,4:3,6-dianhydrohexitols (**1, 20, 21**) have been used as rigid monomers in the synthesis of PU. Polyurethanes containing isosorbide (**1**) have been prepared by several research groups, and complex polyurethanes with elastomeric character and good mechanical properties gave rise to many patents. They were obtained from isosorbide and diisocyanates in the presence of suitable catalysts (**Table 1**). Thus, catalytic polymerization of 2-deoxy-1,4:3,6-dianhydro-2-isocianate-L-iditol (**50**) afforded the corresponding AB-type polyurethane (**Figure 4**). Bachmann et al. [32] described an alternative synthesis through the 2-azido-5-*O*-chloroformyl-1,2-dideoxy-1,4:3,6-dianhydro-L-iditol (**51**) which underwent spontaneous polycondensation, after catalytic hydrogenation, *via* the 2-amino-5-*O*-chloroformyl isoidide. The same authors also described the transformation of the 2,5-diamino-dianhydrohexitols **9, 83,** and **84** into the corresponding diisocyanates **10, 11,** and **13** (**Figure 8**) which was achieved by reaction with phosgene [9]. A dithiodiisocyanate derivative (**12**) was prepared from the 2,5-diamino-2,5-dideoxy-dianhydro-

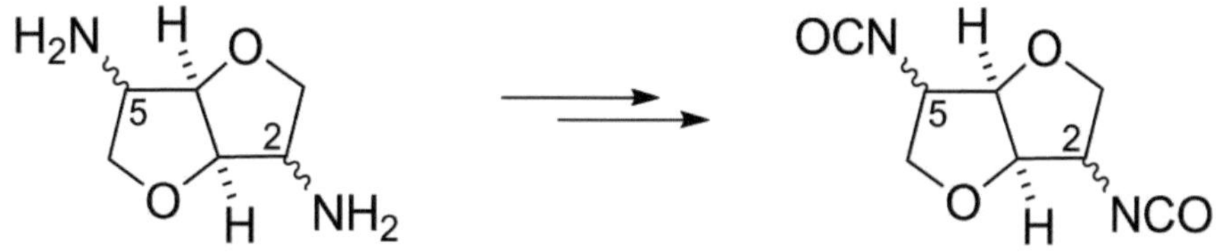

9 D-*Gluco* (2-exo/5-endo)
83 L-*Ido* (2-exo/5-exo)
84 D-*Manno* (2-endo/5-endo)

10 D-*Gluco* (2-exo/5-endo)
11 L-*Ido* (2-exo/5-exo)
13 D-*Manno* (2-endo/5-endo)

Figure 8. Diisocyanates obtained from 2,5-diamino-dianhydrohexitols.

L-iodide (**83**) and thiophosgene, which yielded poly(thio)urethanes and poly(thio)ureas by reaction with the diamino-monomers (**9, 83, 84**).

Beldi et al. [13] studied the polymerization of 1,6-hexamethylene diisocyanate (**3**, HDI) and 4,4'-methylenediphenyl diisocyanate (**23**, MDI) with the 1:4,3:6-dianhydrohexitols (**1, 20, 21**) as well as another novel isosorbide-based ether-diol in DMAc. They established the nature of the end groups and the fraction of cyclic and noncyclic polyurethanes by nuclear magnetic resonance (NMR) and Matrix-Assisted Laser Desorption/Ionization-Time-Of-Flight Mass Spectrometry (MALDI-TOF MS), and demonstrated that the two combined techniques provide a robust method for the identification of structures, chain terminations, and by-products derived from side reactions.

Muñoz-Guerra et al. [18] prepared segmented PU from hydroxyl-end-capped polycaprolactone (3000 g/mol) as soft segment, diisocyanates HDI (**3**) or MDI (**23**) and 1,4-butanediol, isosorbide (**1**), and/or 2,4:3,5-di-*O*-methylene-D-glucitol (**33**) as extenders. The comparative effect of the preparation method (in solution or in bulk) and the influence of the selected extender (**1** or **33**), on the properties of the resulting segmented PU, were evaluated. Hydrolytic degradability was significantly increased by the presence of sugar units, although polymer degradation took place fundamentally by hydrolysis of the polyester soft segment. The same authors also carried out a comparative study of non-segmented polyurethanes with the purpose of evaluating the effect of the replacement of 1,4-butanediol by **1**. The polymerizations were accomplished by standard methods using HDI and MDI as isocyanates [52]. It was observed that T_g values of PU increased with the content in isosorbide.

Koning et al. have investigated water-borne polyurethanes based on isosorbide (**1**) and other renewable building blocks, such as amino acids and fatty acids [26, 27, 53]. The synthesis of water-dispersible polyurethanes prepolymers was carried out from **1**, ethyl ester L-lysine diisocyanate (**46**), dimethylpropionic acid, and a dimer fatty acid-based diisocyanate [53]. The regioselectivities of the endo- and exo-OH functional groups of **1** and the primary ε-NCO and secondary α-NCO of **46** were found to have only minor consequences for the formation of NCO-terminated PU prepolymers. PU dispersions prepared from these four-component prepolymers showed good storage stability. Fully renewable poly(ester urethane urea)s (PEU) were also synthesized from bio-based starting materials: the renewable polyester diol poly(1,2-dimethylethylene adipate), isoidide diisocyanate (**11**), and diaminoisoidide (**83**) as chain extender. The authors found that the PEU based on the isoidide diisocyanate (**11**) exhibited satisfactory thermal and mechanical properties [54].

Thermoplastic and cross-linked bio-based PU with tailored properties and high renewable carbon content were synthesized from isosorbide (**1**), 1,3-propanodiol, and 1,1,1-tris-(hydroxymethyl) propane as the cross-linker reagent, in bulk by varying the molar ratio of the components and without using any chemical catalysts [55].

5. Other conformationally restricted linear PU

The bicyclic structures of the dianhydrohexitols (**1**, **20**, **21**) provide conformational restriction and stiffness to the polymer chain, with a significant increase in the T_g. Galbis et al. [7] reported new conformationally restricted linear polyurethanes based on bicyclic carbohydrate-based monomers having D-*gluco*, *galacto*, and D-*manno* configurations (**33**, **43**, **44**) and their secondary hydroxyl groups protected as cyclic acetals (**Figure 9**). The T_g values of these conformationally restricted polymers were similar to those observed for the polyurethanes [PU(Is-HDI) and PU(Is-MDI)] based on the isosorbide (**1**) [52] and higher than their acyclic analogs. They also proved that the diacetalization of the sugar unit of the polyurethane chain improved the thermal stability, which was comparable to those based on isosorbide.

The two orthoesters **86** and **87** having a rigid adamantane-like structure (**Figure 10**), obtained from naturally occurring *myo*-inositol (**85**), were used as new triol- and diol-type monomers to give the corresponding networked and linear polyurethanes by polyaddition reaction with diisocyanates [56].

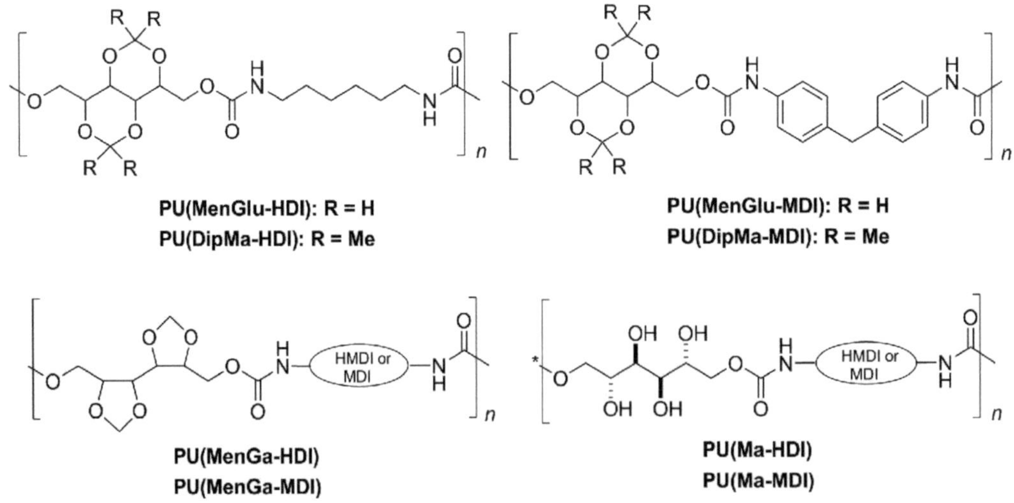

Figure 9. Conformationally restricted hydroxy-polyurethanes based on hexitols.

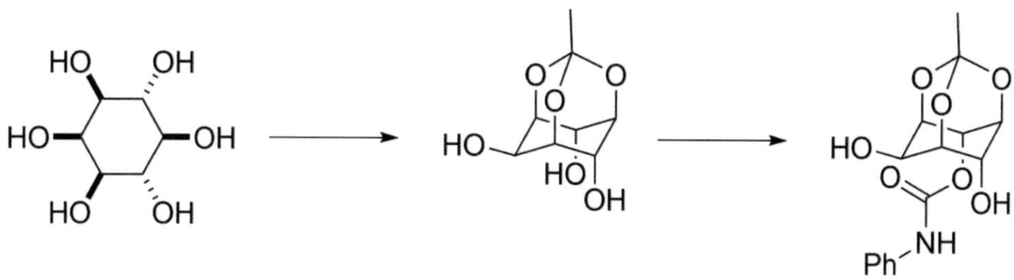

85

86

87

Figure 10. *myo*-Inositol-based rigid adamantine-like monomers.

The linear polyurethanes obtained from **87** showed high values of T_g, ranging from 105 to 177°C, due to the rigidity of the adamantane-like structure units introduced into the main chains. T_g's of the networked polyurethanes obtained from **86** were higher than those of the linear polyurethanes, and ranged from 155 to 248°C.

6. PU with enhanced hydrophilicity

PU with free hydroxyl groups (PHU) have been synthesized from acetalized sugar-based monomers, mainly isopropylidene and benzylidene groups, which are very easily cleaved under acidic conditions. Thus, deacetalization of those polyurethanes containing di-*O*-isopropylidene-D-mannitol units (**44**) yielded multihydroxy polymers in good yields without apparent degradation of the polymer chain (**Figure 9**) [7]. The latter polymers showed enhanced hydrophilicity and hydrolytic degradability as well as lower T_g values and thermal stability than their acetalized counterparts.

Methyl 2,6-di-*O*-pivaloyl-α-D-glucopyranoside (**14**) or methyl 4,6-*O*-benzylidene-α-D-glucopyranoside (**15**) catalyzed by 1,4-diazabicyclo[2.2.2]octane (DABCO) were polymerized with HDI [10]. Likewise, diol **15** was also copolymerized with the methyl ester of L-lysine diisocyanate (**4**) [35] and HDI as well as the diol 1,2:5,6-di-*O*-isopropylidene-D-glucitol (**19**). The corresponding polyhydroxy PUs were obtained by deprotection in aqueous trifluoroacetic acid solution.

A stereoregular polyhydroxy [AB]-polyurethane was prepared from 1-deoxy-1-isocyanate-2,3:4,5-di-*O*-isopropylidene-D-galactitol (**52**) (prepared from 1-amino-1-deoxy-2,3:4,5-di-*O*-isopropylidene-D-galactitol) and the subsequent hydrolysis of the isopropylidene groups (**Figure 11**) [41]. No data on the degradation studies were provided.

Polyaddition reaction of L-gulonic acid-based diols (**16**, **17**) to diisocyanates, followed by the hydrolysis of the isopropylidene groups, yielded PU containing lactone rings and free hydroxyl groups in the main polymer chains (**Figure 12**) [11]. The free hydroxyl groups also enhanced the hydrolysis of lactone rings and, hence, of carbamate groups in the polyurethanes. The multihydroxy polyurethane prepared from L-gulonolactone **16** and HDI was degradable at pH 8.0 under mild temperatures.

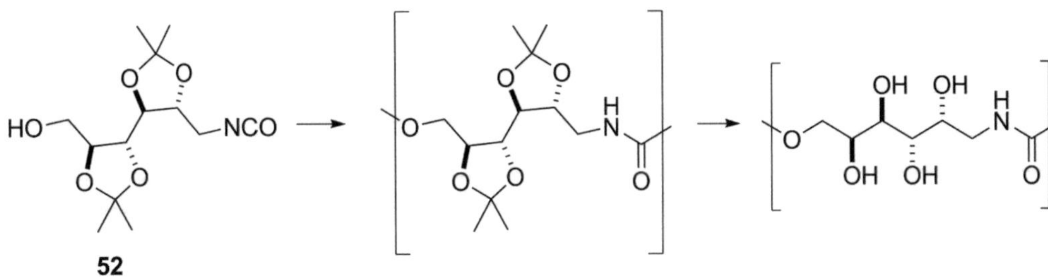

52

Figure 11. [AB]-Polyurethanes from 1-deoxy-1-isocyanate-2,3:4,5-di-*O*-isopropylidene-D-galactitol.

1,2:5,6-di-*O*-Isopropylidene-D-mannitol (**37**) and 1,2-*O*-isopropylidene-D/L-erythritol (**38/39**) were the starting materials for the preparation of new linear multihydroxy polyurethanes by polyaddition with HDI and MELDI (**4**), and subsequent deprotection in acidic media [23]. Likewise, copolyurethanes from **37** and poly(oxytetramethylene) glycol were also prepared to estimate the effects of the D-mannitol unit on their degradability.

The *O*-benzyl derivatives **29, 31,** and **32** with L-*arabino*, L-*threo*, and *xylo* configurations, respectively, were used to prepare PU, and the effect of pendant bulky benzyl groups in the polymer chain was investigated [16, 17]. The removal of benzyl groups was attempted by hydrogenolysis, and the best results were obtained for the PU derived from L-threitol and HDI, which became debenzylated up to 70%. It was found that *O*-benzylated PU were highly resistant to hydrolytic degradation, whereas PU with free hydroxyl groups degraded to a great extent under physiological conditions.

To achieve a facile preparation of sugar-based multihydroxy PU, the use of unprotected saccharides was also investigated. Thiem et al. reported the synthesis of novel PU and polyureas based

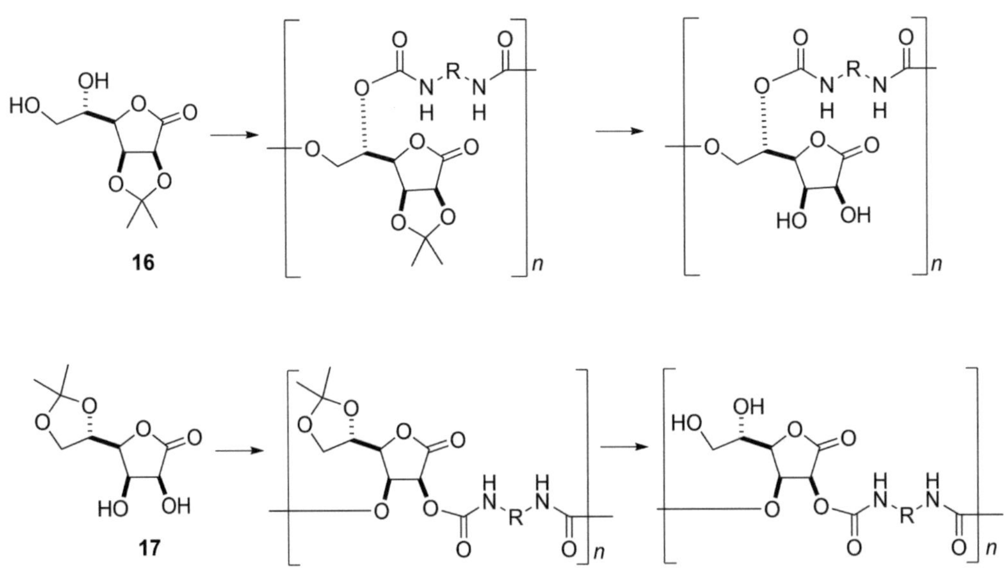

Figure 12. Polyaddition of L-gulonic acid-based diols to diisocyanates.

on modified glycosylamines and glucosamines by catalytic polymerizations [31]. It was found that the anomeric hydroxyl groups were more reactive than the amino groups.

The selective reaction of primary hydroxyl groups of xylitol (**55**) with dimethyl hexamethylene dicarbamate (HDC, **56**) or di-*tert*-butyl-4,4'-diphenyl methyl dicarbamate (MDC, **59**) led to two new linear polyurethanes [PU(X-HDC) and PU(X-MDC)] (**Figure 13**) [6]. Likewise, by the reaction of xylitol with the analogous diisocyanates HDI or MDI, similar polyurethanes [PU(X-HDI) and PU(X-MDI)] were obtained. However, the reaction conditions needed to be adjusted, and so low temperatures were required. Even so, a certain degree of cross-linking was encountered because of the higher reactivity of the diisocyanate comonomers. PU(X-MDC) and PU(X-MDC) were semicrystalline materials showing well-defined melting transitions with high melting enthalpies.

Two novel sugar-based polyol monomers from methyl α-D-glucopyranoside and sucrose and epoxidized methyl oleate were synthesized (**Figure 14**) [51]. Linear and cross-linked PU were obtained by polyaddition with isophorone diisocyanate (**80**) as comonomer, at 60°C using DBTDL as catalyst. The amphiphilic nature of the sugar-based monomers had a marked impact on the final product isolated. Thus, linear or cross-linked PU were obtained depending on the solvent used, that is, DMF or THF. It was found that the polyol monomers were fully soluble in DMF, and therefore cross-linked PU were obtained. By contrast, the formation of linear PU with one pendant sugar moiety per monomer unit was attained in THF. The hydroxyl functions from the sugar moiety were quasi nonreactive under those conditions due to the self-assembly of the sugar-based polyols into nanoparticle structures.

Although polyurethanes are widely investigated, their sulfur analogs, polythiourethanes (PTU), are a relatively poorly investigated group of polymeric materials [57]. The synthesis and characterization of a new linear functional polythiourethane based on D,L-1,4-dithiothreitol [PTU(DTT-HDI)] has been accomplished (**Figure 15**), and its properties as excipient in drug release formulations investigated [58]. This PU with free hydroxyl groups in its structure showed a major ability

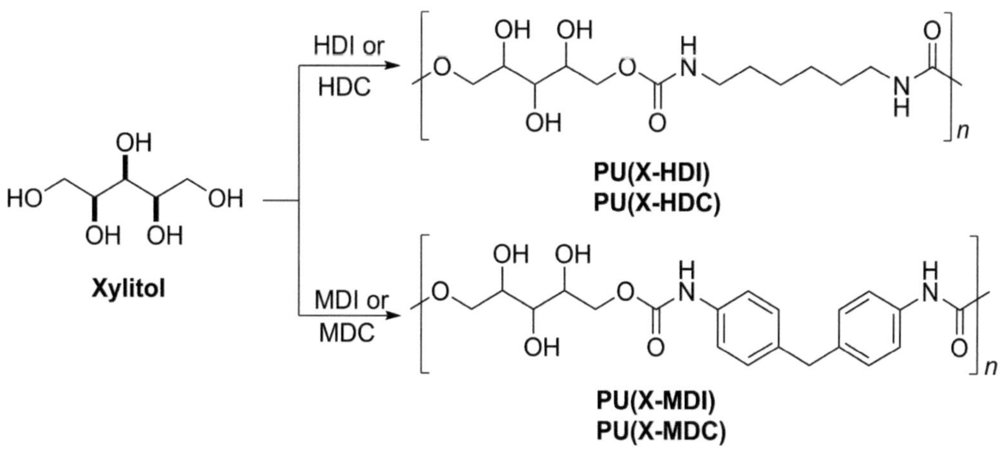

Figure 13. Linear polyurethanes with enhanced hydrophilicity.

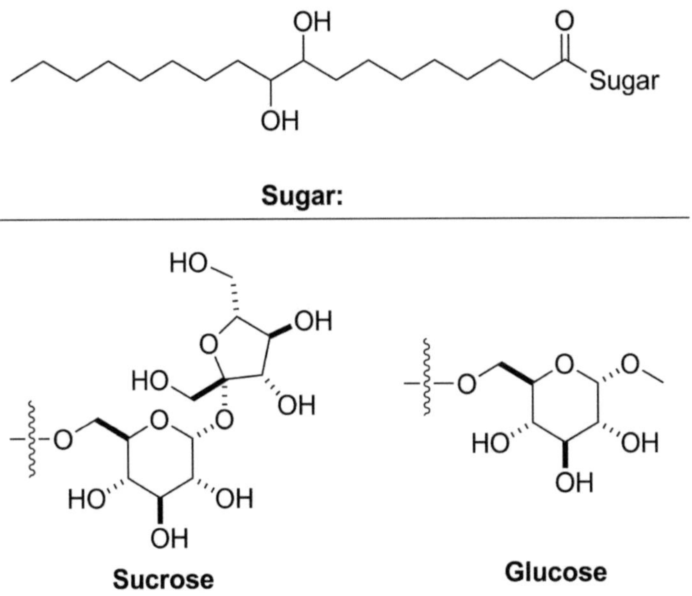

Figure 14. Diol monomers synthesized from methyl a-D-glucopyranoside or sucrose and epoxidized methyl oleate.

to form matrix systems and promoted a significant decrease in the release rate of the model drug theophylline; as a result, it proved to be an excellent controlled release matrix forming excipient.

An homopolyurethane with free secondary hydroxyl groups based on 3,4-*O*-isopropylidene-D-mannitol and 2,2'-dithiodiethyldiisocyanate has been used as sustained matrix forming excipient for site-specific drug release in the gastrointestinal tract [59].

The reaction of diglycerol dicarbonate, synthesized from diglycerol and dimethyl carbonate, and various diamines led to amorphous poly(hydroxy urethane)s, in bulk at mild temperatures, without any catalyst. The abundance of hydroxyl groups along the polymer backbones allows curing purposes and/or further functionalization [60]. Very rigid polyurethane foams with high cross-linking density were obtained from sorbitol-based polyols. The cross-linking density of the formed PU network was directly modified by the polyol mixture ratio, and microstructure and properties also changed in consonance. The incorporation of different amounts of a diol with longer chain length between hydroxyl groups allowed fixing the rigidity of the foams [61].

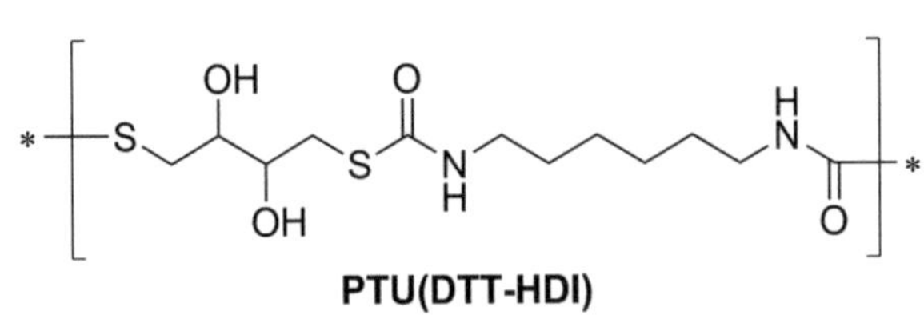

Figure 15. Structure of D,L-1,4-dithiothreitol-based polyurethane.

7. Degradable PU

7.1. Hydrolytic: non-enzyme and enzyme-mediated degradation

The *O*-protected L-threitol, L-arabinitol, and xylitol diols (**24–26, 29, 31, 32**) were used [14] to prepare a series of linear [AABB]-type PU (**Figure 16**) by polyaddition reaction to HDI (**3**) and MDI (**23**). The *O*-methyl-protected polyurethanes derived from **38** and **39** were amorphous and their T_g was highly dependent on the aliphatic or aromatic nature of the diisocyanate used. The hydrodegradability of these PU was conditioned by the number of methoxy side groups present in the repeating unit.

The hydrophilic polyurethanes PU(X-MDC) and PU(X-MDC) (**Figure 13**) obtained by the reaction of xylitol with dimethyl hexamethylene dicarbamate (HDC) or di-*tert*-butyl-4,4'-diphenyl methyl dicarbamate (MDC) [6] were hydrolytically degradable under physiological conditions, in contrast with less-hydrophilic linear polyurethanes previously described [14].

In order to enhance degradability, a series of homo- and copolyurethanes containing the hydrophilic monomer [triethylene glycol (TEG)] and 1,4-di-*S*-benzyl-D,L-dithiothreitol (DTTSBn) (**34**) with HDI (**3**) [19] was synthesized (**Figure 17**). Enzymatic degradation studies were carried out with a variety of proteolytic and esterase enzymes. They were thermally stable up to 250°C, and PU(TEG-HDI) homopolymer was degraded under physiological conditions. In addition, a study of the properties of this PU as matrix-forming excipient for controlled drug delivery was carried out [20].

7.2. Reductive degradation

Galbis et al. reported the introduction of disulfide linkage into the polymer backbone of novel reduction-sensitive biodegradable sugar-based polyurethanes. Although hydrolytical degradation of PU has been extensively reported, a faster degradation method under milder degradation conditions was achieved mediated by glutathione (GSH) [21]. Thus, polyaddition reactions of mixtures of 2,2'-dithiodiethanol (DiT) and 2,3,4-tri-*O*-methyl-L-arabinitol (**25**), 2,3,4-tri-*O*-benzyl-L-arabinitol (**29**) or L-arabinitol (**35**) to HDI (**3**) were carried out (**Figure 18**). PU(DiT-HDI) homopolymer exhibited high crystallinity but the introduction of the L-arabinitol-based diols led to a reduction in the crystallinity of the copolymers. All the copolyurethanes were biodegraded under physiological conditions being crystallinity an important factor in the degradation rate. Thus, the amorphous copolymers with low DiT repeating unit contents were more easily cleaved, despite the lower disulfide ratio. Their DiT repeating units do not pack into semicrystalline segments and

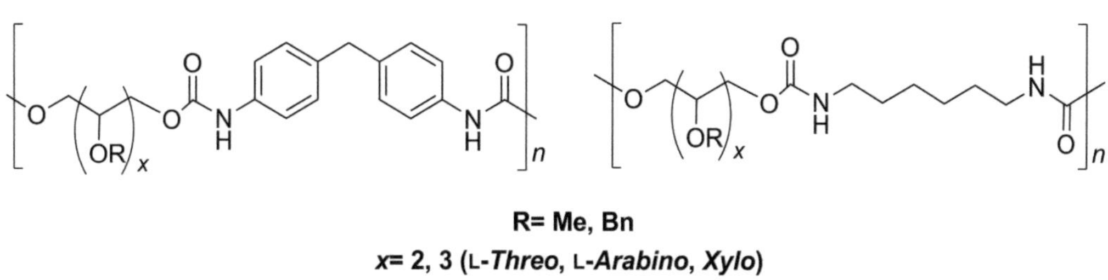

Figure 16. Linear polyurethanes derived from *O*-protected alditols.

were, as a result, more accessible to the reducing agent. Therefore, the amorphous copolymers PU [($ArBn_{80}$-DiT_{20})-HMDI] and PU[($ArMe_{80}$-DiT_{20})-HMDI] were those with enhanced degradation trends.

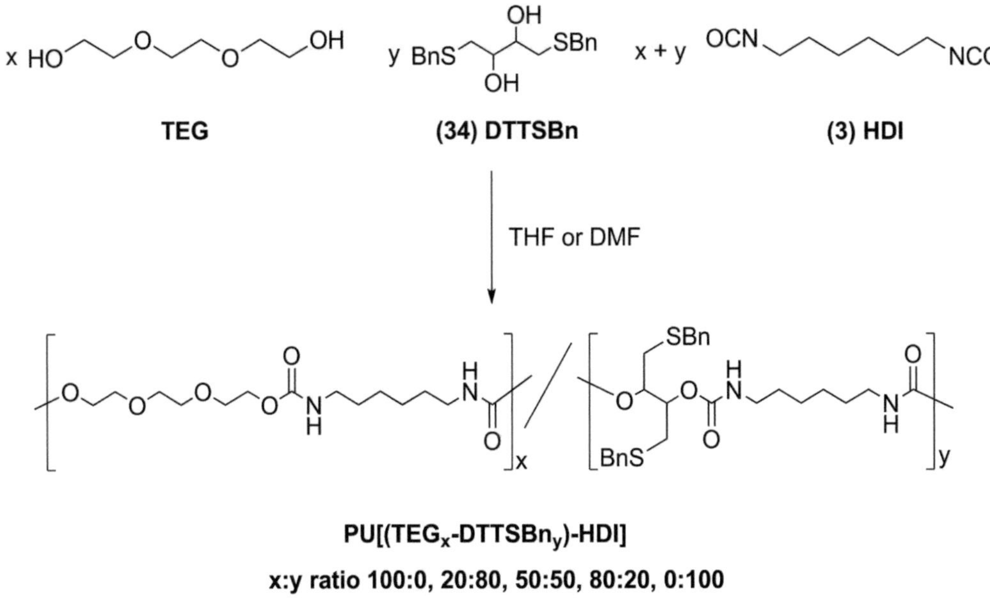

Figure 17. Copolyurethanes derived from 1,4-di-*S*-benzyl-D,L-dithiothreitol.

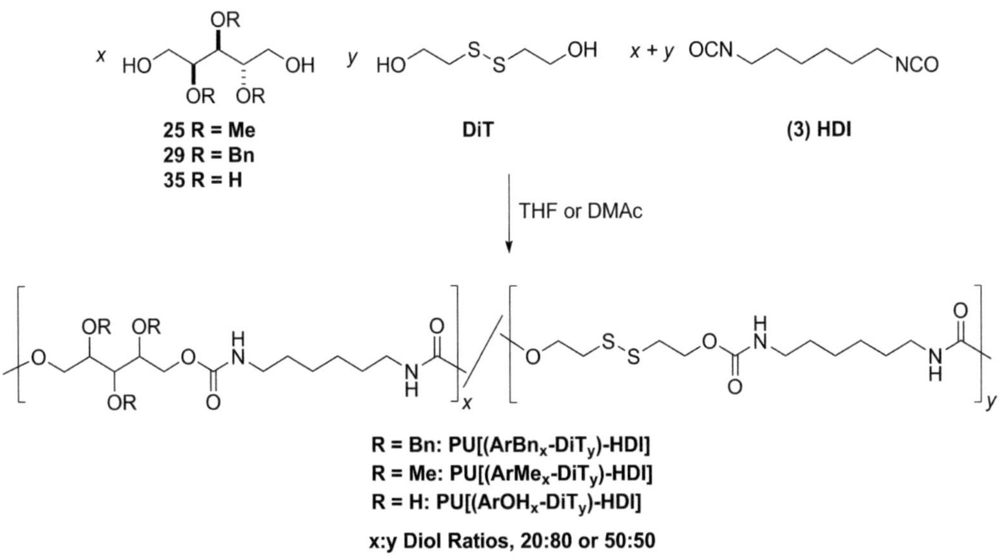

Figure 18. Copolyurethanes degradable by glutathione under physiological conditions.

8. Functional PU

The chemical modification of polymers is an important task for further applications. In this sense, click chemistry (CC) is the most popular approach to achieve chemical transformations; for instance, the thiol-ene coupling reaction of a thiol and an alkene/alkyne and the alkyne-azide cycloaddition reaction.

Galbis et al. described the preparation of new polymerizable diols based on di- and tri-*O*-allyl- and tri-*O*-propargyl-L-arabinitol derivatives prepared from L-arabinitol (**36**, **40**, **49**) as versatile materials for the preparation of tailor-made polyurethanes with varied degrees of functionalization, such as NHBoc, carboxylic or 1,2-dihydroxyethyl side groups (**Figures 19–21**) [22, 24]. This strategy provides a simple and versatile platform for the design of new materials whose functionality can be easily modified to anchor diverse biologically active molecules. From

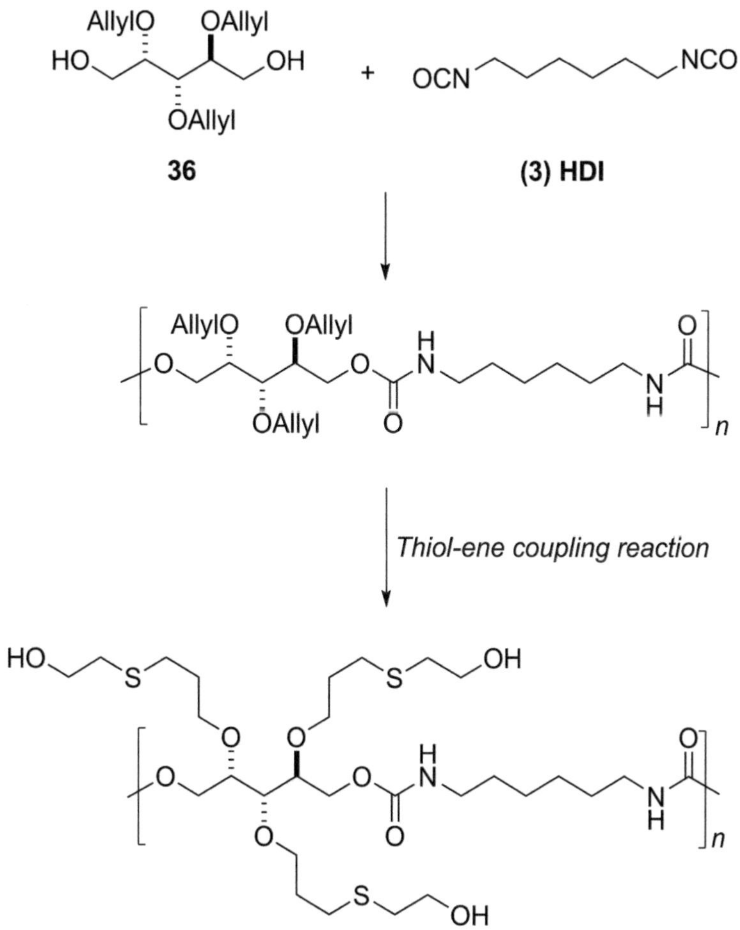

Figure 19. Functionalization of *O*-allyl polyurethanes by the click thiol-ene reaction.

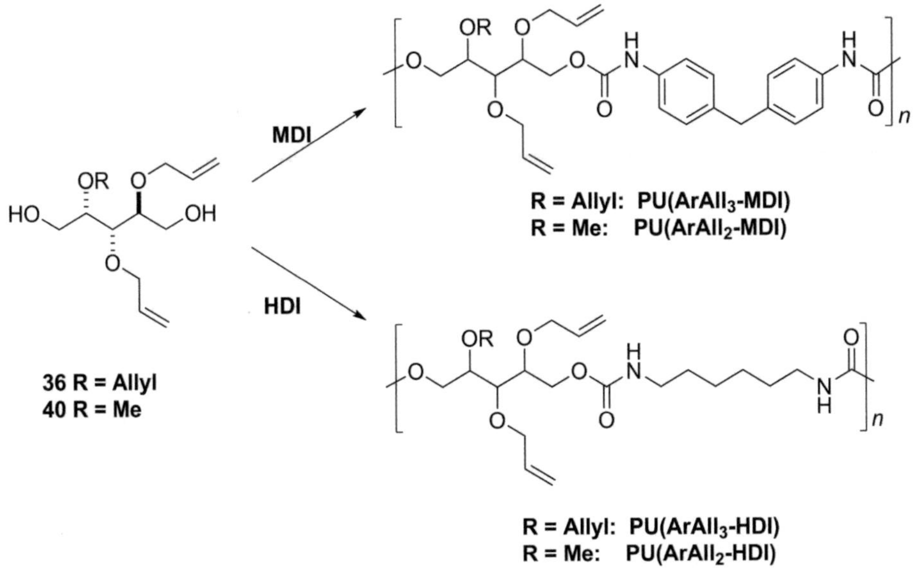

Figure 20. Synthesis of allyl homopolyurethanes.

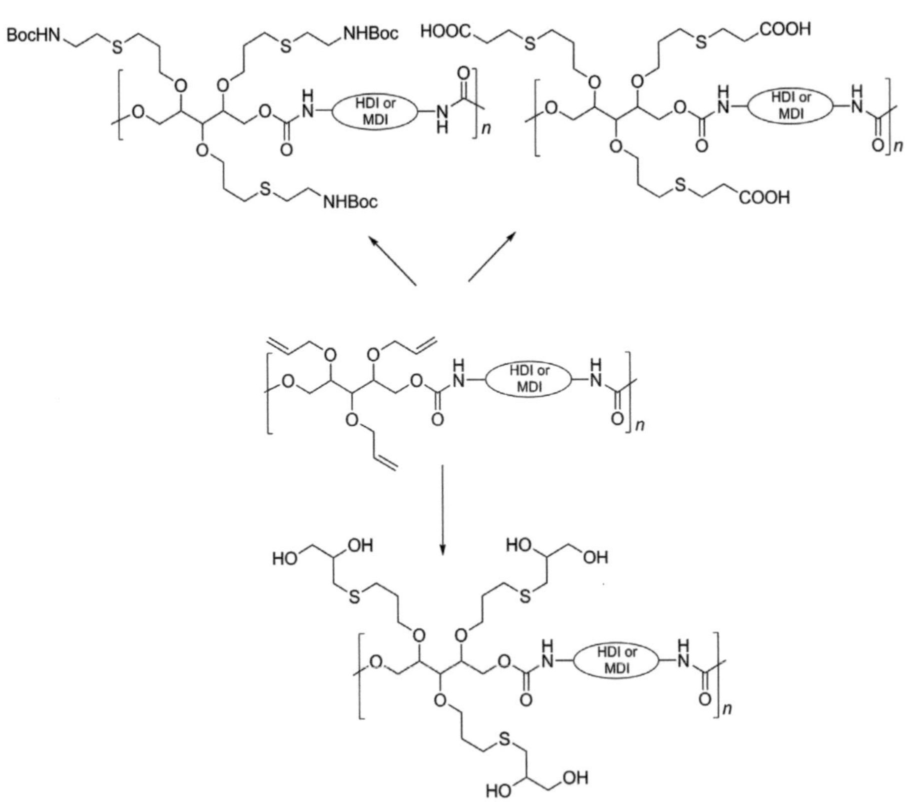

Figure 21. Functional polyurethanes prepared by click reactions on *O*-allyl polyurethanes.

monomer **49,** a series of new biodegradable comb-like copolyurethanes was achieved by the combination of step-growth polymerization and click chemistry [30]. Stiffness, hydrophilic/lipophilic character, and degradability could be adjusted by the appropriate selection of the co-monomers and their feed ratios.

The preparation of novel reduction-sensitive biodegradable multiallyl- and multiamine-based copolyurethanes was also accomplished. The two functional allyl diol monomers **36** and **40** and the commercial 2,2'-dithiodiethanol (DiT) were made to react with HDI (**3**) and MDI (**23**) to give novel linear multiallyl-based copolyurethanes, useful as carriers of anionic drugs or gene materials [25]. Thus, reduction-sensitive multiamine copolymers were obtained [25] from sugar-based diol monomer $Ar(NHBoc)_3$(**41**), with the amine groups conveniently protected to prevent undesirable side reactions (**Figure 22**). Subsequent *N*-deprotection led to the target materials in high yields. The copolymers functionalized with amino groups (protected or free) exhibited lower thermal stability than those bearing allyloxy pendant groups. Functionalization of the sugar moieties greatly affected the T_g of the polymer precursors. At the same time, the DiT diol content controlled the T_m values of the random copolymers. The best degradation rates in reductive environments were obtained for the aliphatic amino-based copolymer, with an associated weight loss close to 90% after 10 days. This polymer was used in the preparation of colon-targeted drug delivery systems. An *in vitro* study of tablets containing the anticancer drug methotrexate demonstrated its impressive ability to control the drug's release, despite its low concentration in the tablet.

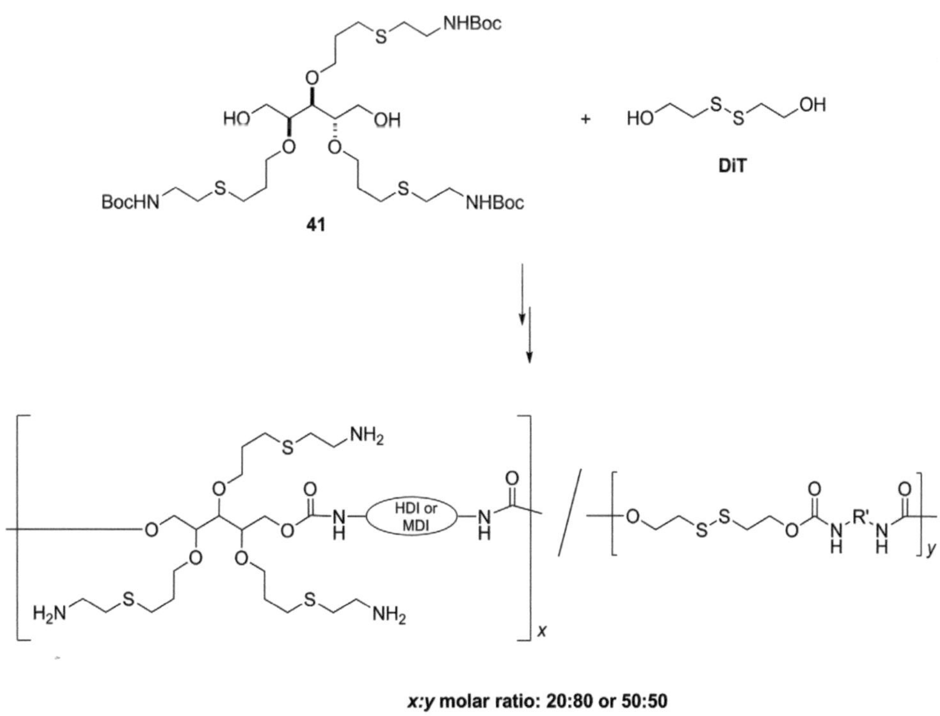

Figure 22. Synthesis of reduction-sensitive multiamine copolyurethanes.

Similarly, the preparation of PU with well-defined "tunable" macromolecular comb-like architectures was successfully accomplished [30] by means of a two-step procedure: preparation of multipropargyl copolyurethanes based on the tri-*O*-propargyl-L-arabinitol derivative **49** and DiT and their subsequent derivatization via alkyne-azide click reaction with chemically diverse azides—dodecyl azide, triethylene glycol azide, and polyethylene glycol azide. This methodology allowed the generation of PU with an extensive chemical diversity that has been unapproachable until now. It is remarkable that a wide variation in degradability under reductive environment could be attained simply by adjusting the chemical structure of the polymer to synthesize.

9. Conclusions

Monosaccharide-based polyurethanes (PU) have been usually prepared by step-growth polymerization from diisocyanates and sugar-based diols but, in the last years, and due to the toxicity of stannous catalysts and common aromatic diisocyanates, other eco-friendly methods have been developed and many works have been reported on the synthesis of isocyanate-free PU, also called non-isocyanate PU (NIPU). Although most syntheses of high-molecular-weight linear polyurethanes involve monosaccharide derivatives having the hydroxyl groups appropriately blocked, several PU with enhanced hydrophilicity and hydrolytic degradability have been synthesized using sugar-derived monomers with free hydroxyl groups or by deprotection of PU prepared from *O*-protected sugar-based monomers.

In order to enhance degradability, a series of homo- and copolyurethanes containing the hydrophilic monomer [triethylene glycol (TEG)] was synthesized. The introduction of disulfide linkage into the polymer backbone of novel reduction-sensitive biodegradable sugar-based polyurethanes has also been reported. Thus, a faster glutathione-mediated degradation method was accomplished under milder degradation conditions.

The chemical modification of PU is important in order to search further applications. Click chemistry (CC) is the most widely used approach to achieve these chemical transformations; for instance, the thiol-ene/yne coupling reaction and the alkyne-azide cycloaddition reaction. In this sense, di- and tri-*O*-allyl- and tri-*O*-propargyl-L-arabinitol derivatives have been used as versatile materials for the preparation of tailor-made polyurethanes with varied degrees of functionalization.

We can conclude that it is viable to prepare monosaccharide-based polyurethanes analogous to the more qualified conventional polymers, but having a larger functionality and an enhanced hydrophilicity and degradability. However, in most cases, the high costs that represent the preparation of the monomers restrict the application of these polymers to the biomedical and other specialized fields.

Author details

Juan A. Galbis*, María de Gracia García-Martín, María Violante de Paz and Elsa Galbis

*Address all correspondence to: jgalbis@us.es

Departamento de Química Orgánica y Farmacéutica, Universidad de Sevilla, Seville, Spain

References

[1] Gandini A, Lacerda TM, Carvalho AJF, Trovatti E. Progress of polymers from renewable resources: Furans, vegetable oils and polysaccharides. Chemical Reviews. 2016;**116**:1637–1669. DOI: 10.1021/acs.chemrev.5b00264

[2] Gandini A, Lacerda TM. From monomers to polymers from renewable resources: Recent advances. Progress in Polymer Science. 2015;**48**:1–39. DOI: 10.1016/j.progpolymsci.201411.002

[3] Galbis JA, García-Martín MG, de Paz MV, Galbis E. Synthetic polymers from sugar-based monomers. Chemical Reviews. 2016;**116**:1600–1636. DOI: 10.1021/acs.chemrev.5b00242

[4] Kiely DE, Chen L, Lin TH. Synthetic polyhydroxypoliamides from galactaric, xylaric, D-glucaric, and D-mannaric acids and alkylenediamine monomers. Some comparisons. Journal of Polymer Science Part A. Polymer Chemistry. 2000;**38**:594–603. DOI: 10.1002/(SICI)1099-0518(20000201)38:3<594::AID-POLA24>3.0.CO;2-#

[5] Styron SD, Kiely DE, Ponder G. Alternating stereoregular head, tail-tail, head-poly (alkylene D-glucaramides derived from a homogeneous series of symmetrical diamido-di-D-glucaric acid monomers. Journal of Carbohydrate Chemistry. 2003;**22**:123–142. DOI: 10.1081/CAR-120020482

[6] Begines B, Zamora F, Roffé I, Mancera M, Galbis JA. Sugar-based hydrophilic polyurethanes and polyureas. Journal of Polymer Science Part A. Polymer Chemistry. 2011;**49**:1953–1961. DOI: 10.1002/pola.24621

[7] Begines B, Zamora F, Benito E, Garcia-Martin MG, Galbis JA. Conformationally restricted linear polyurethanes from acetalized sugar-based monomers. Journal of Polymer Science Part A. Polymer Chemistry. 2012;**50**:4638–4646. DOI: 10.1002/pola.26293

[8] Wilbullucksanakul S, Hashimoto K, Okada M. Synthesis of polyurethanes from saccharide-derived diols and diisocyanates and their hydrolyzability. Macromolecular Chemistry and Physics. 1996;**197**:135–146. DOI: 10.1002/macp.1996.021970110

[9] Bachmann F, Reimer J, Ruppenstein M, Thiem J. Mint: Synthesis of novel polyurethanes and polyureas by polyaddition reactions of dianhydrohexitol configurated diisocyanates. Macromolecular Chemistry and Physics. 2001;**202**:3410–3419. DOI: 10.1002/1521-3935(20011101)202:17<3410::AID-MACP3410>3.0.CO;2-Q

[10] Garcon R, Clerk C, Gesson JP, Bordado J, Nunes T, Caroco S, Gomes PT, Da Piedade MEM, Rauter AP. Synthesis of novel polyurethanes from sugars and 1,6-hexamethylene diisocyanate. Carbohydrate Polymers. 2001;**45**:123–127. DOI: 101016/S0144-8617(00)00323-4

[11] Yamanaka C, Hashimoto K. Synthesis of new hydrolyzable polyurethanes from L-gulonic acid-derived diols and diisocyanates. Journal of Polymer Science Part A. Polymer Chemistry. 2002;**40**:4158–4166. DOI: 10.1002/pola.10496

[12] Hashimoto K, Yaginuma K, Nara S, Okawa H. Synthesis of new hydroxy-bearing polyurethanes: Polyaddition of D-glucose-derived diols with diisocyanates. Polymer Journal. 2005;**37**:384–390. DOI: 10.1295/polymj.37.384

[13] Beldi M, Medimagh R, Chatti S, Marque S, Prim D, Loupy A, Delolme F. Characterization of cyclic and non-cyclic poly-(ether-urethane)s bio-based sugar diols by a combination of MALDI-TOF and NMR. European Polymer Journal. 2007;**43**:3415–3433. DOI: 10.1016/j.eurpolymj.2007.06.003

[14] de Paz MV, Marin R, Zamora F, Hakkou K, Alla A, Galbis JA, Munoz-Guerra, S. Linear polyurethanes derived from alditols and diisocyanates. Journal of Polymer Science Part A. Polymer Chemistry. 2007;**45**:4109–4117. DOI: 10.1002/pola.22127

[15] Marin R, Munoz-Guerra S. Mint: Linear polyurethanes made from threitol: Acetalized and hydroxylated polymers. Journal of Polymer Science Part A. Polymer Chemistry. 2008;**46**:7996–8012. DOI: 10.1002/pola.23099

[16] de Paz MV, Aznar JA, Galbis JA. Versatile sugar derivatives for the synthesis of potential degradable hydrophilic-hydrophobic polyurethanes and polyureas. Journal of Carbohydrate Chemistry. 2008;**27**:120–140. DOI: 10.1080/07328300802030878

[17] Marin R, de Paz MV, Ittobane N, Galbis JA, Munoz-Guerra S. Hydroxylated linear polyurethanes derived from sugar alditols. Macromolecular Chemistry and Physics. 2009;**210**:486–494. DOI: 10.1002/macp.200800517

[18] Marin R, Munoz-Guerra S. Carbohydrate-based poly(ester-urethane)s: A comparative study regarding cyclic alditols extenders and polymerization procedures. Journal of Applied Polymer Science. 2009;**114**:3723–3736. DOI: 10.1002/app.30924

[19] Ferris C, de Paz MV, Zamora F, Galbis JA. Dithiothreitol-based polyurethanes. Synthesis and degradation studies. Polymer Degradation and Stability. 2010;**95**:1480–1487. DOI: 10.1016/j.polymdegradstab.2010.06.021

[20] Campiñez MD, Aguilar-de-Leyva A, Ferris C, de Paz MV, Galbis JA, Caraballo I. Study of the properties of the new biodegradable polyurethane PU (TEG-HMDI) as matrix forming excipient for controlled drug delivery. Drug Development and Industrial Pharmacy. 2013;**39**:1758–1764. DOI: 10.3109/03639045.2012.736516

[21] de Paz MV, Zamora F, Begines B, Ferris C, Galbis JA. Glutathione-mediated biodegradable polyurethanes derived from L-arabinitol. Biomacromolecules. 2010;**11**:269–276. DOI: 10.1021/bm9011216

[22] Ferris C, de Paz MV, Galbis JA. L-Arabinitol-based functional polyurethanes. Journal of Polymer Science Part A. Polymer Chemistry. 2011;**49**:1147–1154. DOI: 10.1002/pola.24530

[23] Hashimoto K, Hashimoto N, Kamaya T, Yoshioka J, Okawa H. Synthesis and properties of bio-based polyurethanes bearing hydroxy groups derived from alditols. Journal of Polymer Science Part A. Polymer Chemistry. 2011;**49**:976–985. DOI: 10.1002/pola.24510

[24] Ferris C, de Paz MV, Galbis JA. Synthesis of functional sugar-based polyurethanes. Macromolecular Chemistry and Physics. 2012;**213**:480–488. DOI: 10.1002/macp.201100672

[25] Ferris C, de Paz MV, Aguilar-de-Leyva A, Caraballo I, Galbis JA. Reduction-sensitive functionalized copolyurethanes for biomedical applications. Polymer Chemistry. 2014;**5**:2370–2381. DOI: 10.1039/c3py01572f

[26] Li Y, Noordover BAJ, van Benthem RATM, Koning CE. Chain extension of dimer fatty acid- and sugar-based polyurethanes in aqueous dispersions. European Polymer Journal. 2014;**52**:12–22. DOI: 10.1016/j.eurpolymj.2013.12.007

[27] Li Y, Noordover BAJ, van Benthem RATM, Koning CE. Property profile of poly(urethane urea) dispersions containing dimer fatty acid-, sugar- and amino acid-based building blocks. European Polymer Journal. 2014;**59**:8–18. DOI: 10.1016/j.eurpolymj.2014.06.016

[28] Zenner MD, Xia Y, Chen JS, Kessler MR. Mint: Polyurethanes from isosorbide-based diisocyanates. ChemSusChem. 2013;**6**:1182–1185. DOI: 10.1002/cssc.201300126

[29] Zenner MD, Madbouly SA, Chen JS, Kessler MR. Mint: Unexpected tackifiers from isosorbide. ChemSusChem. 2015;**8**:448–451. DOI: 10.1002/cssc.201402667

[30] Begines B, de-Paz MV, Alcudia A, Galbis JA. Synthesis of reduction sensitive comb-like polyurethanes using click chemistry. Journal of Polymer Science Part A. Polymer Chemistry. 2016;**54**:3888–3900. DOI: 10.1002/pola.28367

[31] BachmannF, Ruppenstein M, Thiem J. Synthesis of aminosaccharide-derived polymers with urea, urethane, and amide linkages. Journal of Polymer Science Part A. Polymer Chemistry. 2001;**39**:2332–2341. DOI: 10.1002/pola.1210

[32] Bachmann F, Reimer J, Ruppenstein M, Thiem J. Synthesis of a novel starch-derived AB-type polyurethane. Macromolecular Rapid Communications. 1998;**19**:21–26. DOI: 10.1002/(SICI)1521–3927(19980101)19:1<21::AID-MARC21>3.0.CO;2-T

[33] Ionescu M, Petrovic ZS. High functionality polyether polyols based on polyglycerol. Journal of Cellular Plasticity. 2010;**46**:223–237. DOI: 10.1177/0021955X09355887

[34] Balcioglu S, Parlakpinar H, Vardi N, Denkbas EB, Karaaslan MG, Gulgen S, Taslidere E, Koytepe S, Ates B. Design of xylose-based semisynthetic polyurethane tissue adhesives with enhanced bioactivity properties. ACS Applied Materials Interfaces. 2016;**8**:4456–4466. DOI:10.1021/acsami.5b12279

[35] Wibullucksanakul S, Hashimoto K, Okada M. Swelling behavior and controlled release of new hydrolyzablepoly(ether urethane) gels derived from saccharide and L-lysine derivatives and poly(ethylene glycol). Macromolecular Chemistry and Physics. 1996;**197**:1865–1876. DOI: 10.1002/macp.1996.021970608

[36] Ates B, Koytepe S, Karaaslan MG, Balcioglu S, Gulgen S. Biodegradable non-aromatic adhesive polyurethanes based on disaccharides for medical applications. International Journal of Adhesion and Adhesives. 2014;**49**:90–96. DOI: 10.1016/j.ijadhadh.2013.12.012

[37] Pohjanlehto H, Setaelae HM, Kiely DE, McDonald AG. Lignin-xylaric acid-polyurethane-based polymer network systems: Preparation and characterization. Journal of Applied Polymer Science. 2014;**131**:39714–39714. DOI: 10.1002/app.39714

[38] Lalwani R, Desai S. Sorption behavior of biodegradable polyurethanes with carbohydrate crosslinkers. Journal of Applied Polymer Science. 2010;**115**:1296–1305. DOI: 10.1002/app.30214

[39] Solanki A, Mehta J, Thakore S. Structure-property relationships and biocompatibility of carbohydrate crosslinked polyurethanes. Carbohydrate Polymers 2014;**110**:338–344. DOI: 10.1016/j.carbpol.2014.04.021

[40] Kloss J, Bugay C, Akcelrud L, Zawadzki SF, Wang SH. Correlação entre propriedades mecânicas e parâmetros estruturais de poliuretanos à base de poli(ε-caprolactona). Polímeros Ciência e Tacnologia 2005;**15**:1–5. Available from: http://www.redalyc.org/articulo.oa?id=47015103

[41] Gomez RV, Varela O. Synthesis of polyhydroxy [n]-polyurethanes derived from a carbohydrate precursor. Macromolecules. 2009;**42**:8112–8117. DOI: 10.1021/ma9014643

[42] Kolender AA, Arce SM, Varela O. Synthesis and characterization of poly-O-methyl-[n]-polyurethane from a D-glucamine-based monomer. Carbohydrate Research. 2011;**346**: 1398–1405. DOI: 10.1016/j.carres.2011.02.022

[43] Maisonneuve L, Lamarzelle O, Rix E, Grau E, Cramail H. Isocyanate-free routes to polyurethanes and poly(hydroxy urethane)s. Chemical Review. 2015;**115**:12407–12439. DOI: 10.1021/acs.chemrev.5b00355

[44] Cornille A, Auvergne R, Figovsky O, Boutevin B, Caillol S. A perspective approach to sustainable routes for non-isocyanate polyur-ethanes. European Polymer Journal. 2017;**87**:535–552. DOI:10.1016/j.eurpolymj.2016.11.027

[45] Prömpers G, Keul H, Höcker H. Polyurethanes with pendant hydroxy groups: polycondensation of D-mannitol-1,2:5,6-dicarbonate with diamines. Designed Monomers and Polymers. 2005;**8**(6):547–569. DOI: 10.1163/156855505774597830

[46] Fidalgo DM, Kolender AA, Varela O. Stereoregular poly-O-methyl [m,n]-polyurethanes derived from D-mannitol. Journal of Polymer Science Part A. Polymer Chemistry. 2013;**51**:463–470. DOI:10.1002/pola.26406

[47] Besse V, Auvergne R, Carlotti S, Boutevin G, Otazaghine B, Caillol S, Pascault JP, Boutevin B. Synthesis of isosorbide based polyurethanes: An isocyanate free method. Reactive and Functional Polymers. 2013;**73**(3):588–594. DOI: 10.1016/j.reactfunctpolym.2013.01.002

[48] Begines B, Zamora F, de Paz MV, Hakkou K, Galbis JA. Polyurethanes derived from carbohydrates and cystine-based monomers. Journal of Applied Polymer Science. 2015;**132**:41304/1-41304/8. DOI: 10.1002/app.41304

[49] Brocas A-L, Cendejas G, Caillol S, Deffieux A, Carlotti S. Controlled synthesis of polyepichlorohydrin with pendant cyclic carbonate functions for isocyanate-free polyurethane networks. Journal of Polymer Science Part A. Polymer Chemistry. 2011;**49**(12):2677–2684. DOI: 10.1002/pola.24699

[50] Kloss J, de Souza FSM, da Silva ER, Dionisio JA, Akcelrud L, Zawadzki SF. Polyurethanes elastomers based on poly(ε-caprolactone) diol: Biodegradation evaluation. Macromolecular Symposium. 2006;**245/246**:651–656. DOI: 10.1002/masy.200651392

[51] Boyer A, Lingome CE, Condassamy O, Schappacher M, Moebs-Sanchez S, Queneau Y, Gadenne B, Alfos C, Cramail H. Glycolipids as a source of polyols for the design of original linear and cross-linked polyurethanes. Polymer Chemistry. 2013;**4**:296–306. DOI: 10.1039/c2py20588b

[52] Marin R, Alla A, Martinez de Ilarduya A, Munoz-Guerra S. Carbohydrate-based polyurethanes: A comparative study of polymers made from isosorbide and 1,4-butanediol. Journal of Applied Polymer Science. 2012;**123**:986–994. DOI: 10.1002/app.34545

[53] Li Y, Noordover BAJ, van Benthem RATM, Koning CE. Reactivity and regio-selectivity of renewable building blocks for the synthesis of water-dispersible polyurethane prepolymers. ACS Sustainable Chemistry & Engineering. 2014;**2**:788–797. DOI: 10.1021/sc400459q

[54] Tang D, Thiyagarajan S, Noordover BAJ, Koning CE, van Es DS, van Haveren J. Fully renewable thermoplastic poly(ester urethane urea)s from bio-based diisocyanates. Journal of Renewable Materials. 2013;**1**:222–229. DOI: 10.7569/JRM.2013.634116

[55] Calvo-Correas T, Gabilondo N, Alonso-Varona A, Palomares T. Synthesis and characterization of polyurethanes with high renewable carbon content and tailored properties. ACS Sustainable Chemistry & Engineering. 2016;**4**:5684–5692. DOI: 10.1016/j.eurpolymj.2016.03.030

[56] Okamoto S, Onoue S, Kobayashi M, Sudo A. Rigid triol and diol with adamantane-like core derived from naturally occurring *myo*-inositol and their polyaddition with diisocyanates. Journal of Polymer Science Part A. Polymer Chemistry. 2014;**52**:3498–3505. DOI: 10.1002/pola.27414

[57] Kultys A, Rogulska M, Pikus S. The synthesis and characterization of new thermoplastic poly(thiourethane-urethane)s. Journal of Polymer Science Part A. Polymer Chemistry. 2008;**46**:1770–1782. DOI: 10.1002/pola.22520

[58] Campiñez MD, Ferris C, de Paz MV, Aguilar-de-Leyva A, Galbis JA, Caraballo I. A new biodegradable polythiourethane as controlled release matrix polymer. International Journal of Pharmaceutical. 2015;**480**:63–72. DOI: 10.1016/j.ijpharm.2015.01.011

[59] Campiñez MD, Benito E, Romero-Azogil L, Aguilar-de-Leyva A, García-Martín MG, Galbis JA, Caraballo I. Development and characterization of new functionalized polyurethanes for sustained and site-specific drug release in the gastrointestinal tract. European Journal of Pharmaceutical Science. 2017;**100**:285–295. DOI: 10.1016/j.ejps.2017.01.017

[60] Van Velthoven JLJ, Gootjes L, van Es DS, Noordover BAJ. Poly(hydroxy urethane)s based on renewable diglycerol dicarbonate. European Polymer Journal. 2015;**70**:125–135. DOI: 10.1016/j.eurpolymj.2015.07.011

[61] Ugarte L, Gómez-Fernández S, Pen~a-Rodríguez C, Prociak A, Corcuera MA, Eceiza A. Tailoring mechanical properties of rigid polyurethane foams by sorbitol and corn derived biopolyol mixtures. ACS Sustainable Chemistry & Engineering. 2015;**3**:3382–3387. DOI: 10.1021/acssuschemeng.5b01094

Chapter 8

Dynamic Mechanical Behaviour of Polymer Materials

Jitang Fan, Xiaoyun Fan and Ang Chen

Additional information is available at the end of the chapter

http://dx.doi.org/10.5772/intechopen.69570

Abstract

In this chapter, we present the current research program of dynamic mechanical behaviour of polymer materials. Polymers typically have low strength, stiffness and wave impedance, making it challenging to obtain accurate stress-strain relations at high strain rates. The challenges in split Hopkinson bar testing technique modified for characterizing the mechanical response of polymers under high-rate loading are introduced first. Then, various representative polymers are reviewed for illustrating the strain-rate dependencies of mechanical data and their constitutive modellings.

Keywords: polymer, strain rate, dynamic mechanical response

1. Introduction

Exploring the potentials of advanced materials and structures for serving in extreme conditions has been a hot research topic for decades [1–5]. Materials and material technologies can be tailored nowadays for the applications in structural engineering. In recent years, impact-resistant devices and structures have received considerable attention because of the engineering requirements of safety and security. They require the dedicated design concepts and optimized material properties. Most materials currently available lead to the increase in weight and/or volume of structures and devices for meeting the impact-resistant requirements. In many cases, these increases are unacceptable in structure design. For solving this problem, the trend in materials is towards weight reduction and strength increase at reasonable costs. Material technologies are enablers to meet these requirements. They provide the potential to design advanced materials with high special strength (strength-to-weight ratio) and toughness for the applications in dynamic events.

Polymer materials have become a widely concerned research focus in extreme conditions because of their outstanding performances, such as impact resistance, rate dependency, corrosion resistance, low density, easy moulding and so on. Even more, in some engineering applications, they show outstanding advantages and replace the traditional metals and nonmetallic materials [1, 2].

Therefore, understanding the dynamic mechanical behaviour of polymers is necessary. Especially, the investigations of polymers in explosion, shock, collision and other related mechanical behaviour under impact loading have both theoretical and practical significances. In this chapter, the current research program of dynamic mechanical behaviour of various representative polymers is presented.

2. Characterization technology

Due to the action of high strain-rate loading [6–10], dynamic mechanical response of materials is different from that under static loading. Characterizing dynamic mechanical behaviour is full of interest for material applications in impact events. For polymers, because of their low wave impedance, a longer time is needed to reach the dynamic stress equilibrium during split Hopkinson bar (SHB) test [11, 12], which produces more difficulties in high-rate mechanical experiments.

In order to reduce the difference in wave impedance, the viscoelastic impact bar is used in SHB test [13]. However, the viscoelastic bar itself has a strain-rate sensitivity, which easily results in the dispersion and attenuation of stress wave during the travelling in bar [14]. Aluminium alloy bar seems to be effective for testing polymer materials, thanks to the match of wave impedance and its lower strain-rate sensitivity. For meeting the conditions of constant strain rate and dynamic stress equilibrium, it is necessary to accurately design the shape of incident pulse [15]. A pulse-shaping technique is applied to meet these requirements, which, in general, places a wafer at the end of incident bar. The wafer material can be copper. The advantage of this pulse-shaping technique is to reduce the oscillation of incident wave. By selecting the appropriate specimen shape and size, the constant strain-rate loading with uniform stress in specimen can be realized [8, 16, 17]. Besides, low wave impedance material in SHB experiment will make the transmitter signal weak, which results in a small difference between the incident signal and reflected signal. The semiconductor strain gauge is more sensitive [18], which can be used to detect the weak signal. So, it can be used in the SHB experiment to test the soft polymer materials.

Moreover, it is necessary to clarify the calculating methods of stress, strain and strain rate of the tested specimen from the signals recorded by strain gages in an SHB test. They are based on the assumption of one-dimensional stress wave theory. Two long slender elastic bars in SHB apparatus are shown in **Figure 1**, which sandwich a short cylindrical specimen in between them.

When the striker bar impacts the end of the incident bar, a compressive stress wave is produced that immediately begins to travel along the bar towards the specimen. Upon arriving at the specimen, the compressive stress wave partially reflects back towards the impact end of the incident bar. The remainder of the compressive stress wave transmits through the specimen

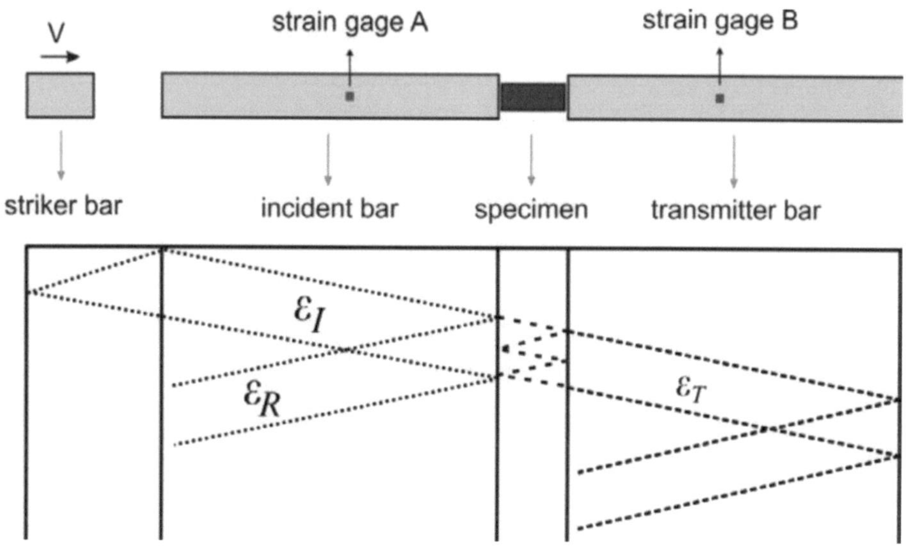

Figure 1. Schematic representation of the elastic waves travelling in a split Hopkinson bar (SHB) apparatus [8].

and enters into the transmitter bar, resulting in the deformation of the tested specimen [8]. Herein, the acoustic impedance of the specimen should be lower than the bar's impedance. The reflected pulse is a tensional wave, whereas the transmitted pulse remains in compression. The stress wave-induced strain histories of the incident bar and transmitter bar are recorded by strain gages A and B, respectively.

As long as the stress in the two bars remains under their elastic limits, specimen stress, strain and strain rate may be calculated from the recorded strain histories of the incident bar and transmitter bar. Herein, two important strain histories are needed to be identified, which are the reflected wave and the transmitted wave through the specimen. Kolsky [7, 8] applied one-dimensional stress wave theory to develop the following relation for calculating the specimen's engineering stress, $\sigma_S(t)$:

$$\sigma_S(t) = E\frac{A_0}{A_S}\varepsilon_T(t) \tag{1}$$

in which E is the transmitter bar's elastic modulus; A_0 is the transmitter bar's cross-sectional area; A_S is the specimen's cross-sectional area before loading and $\varepsilon_T(t)$ is the transmitted strain history.

The specimen strain rate, $\acute{\varepsilon}_S(t)$ can be calculated as:

$$\acute{\varepsilon}_S(t) = \frac{d\varepsilon_S(t)}{dt} = \frac{-2C_0}{L}\varepsilon_R(t) \tag{2}$$

in which $\varepsilon_R(t)$ is the reflected strain history; L is the specimen length prior to impact and C_0 is the wave velocity in the incident bar. The reflected wave represents the strain-rate history in the specimen. So, the flat plateau in the reflected signal corresponds to a nearly constant rate of the specimen deformation during dynamic loading. The wave velocity in the incident bar, C_0, can be calculated from the theory of elementary vibrations, as below:

$$C_0 = \sqrt{E/\rho} \quad (3)$$

in which E and ρ are the incident bar's elastic modulus and density, respectively. Equation (2) can be integrated with time to attain the specimen's engineering strain history, $\varepsilon_S(t)$, as below:

$$\varepsilon_S(t) = \frac{-2C_0}{L} \int_0^t \varepsilon_R(t) \quad (4)$$

Thus, the data of specimen engineering stress, strain and strain rate can be derived from the recorded strain gage signals in an SHB experiment.

However, a uniform deformation of the specimen during SHB experiment should be maintained so that the experimental results can be clearly documented and interpreted for characterizing the dynamic stress-strain relation of a material.

Dynamic stress equilibrium is important for the validity of an SHB experiment. The stress in the specimen can be expressed in terms of the force exerted on each end surface of the specimen. When the specimen is sandwiched in between the incident bar and transmitter bar under dynamic compressive loading, the forces, $F_1(t)$ and $F_2(t)$, are imposed on the specimen with a diameter of D_S and a length of L_S. The average force, F_{avg}, applied on the specimen can be given as:

$$F_{avg}(t) = \frac{F_1(t) + F_2(t)}{2} \quad (5)$$

Whereas, the average engineering stress on the tested cylindrical specimen is:

$$\sigma_{avg}(t) = \frac{F_{avg}(t)}{\frac{1}{4}\pi D_S{}^2} \quad (6)$$

Herein, the forces, $F_1(t)$ and $F_2(t)$ applied on the specimen end surfaces are from the incident bar and transmitter bar, respectively. These two forces can be expressed in terms of elastic strains in the incident bar and transmitter bar, respectively, and can be calculated from the incident, reflected and transmitted signals, as below:

$$F_1(t) = E[\varepsilon_I(t) + \varepsilon_R(t)]\frac{\pi D_{bar}^2}{4} \quad (7)$$

$$F_2(t) = E\varepsilon_T(t)\frac{\pi D_{bar}^2}{4} \quad (8)$$

in which D_{bar} is the diameter of the incident bar (and the transmitter bar) and $\varepsilon_I(t)$, $\varepsilon_R(t)$ and $\varepsilon_T(t)$ are the strain histories of the incident, reflected and transmitted waves, respectively. For a specimen under high-speed loading, the dynamic stress equilibrium should be met for a constant strain rate, which means that $F_1(t)$ should be equal to $F_2(t)$, as below:

$$F_1(t) = F_2(t) \tag{9}$$

However, in a real SHB experiment, the deviation exists between these two force histories. In order to analyse and quantify the deviation, a criterion for stress equilibrium is employed to compare the forces (or stresses) exerted on the specimen end surfaces. Ravichandran and Subhash [19] introduced a parameter, $R(t)$, to evaluate the proximity to stress equilibrium in the specimen, as below:

$$R(t) = \left|\frac{\Delta\sigma(t)}{\sigma_{avg}(t)}\right| = \left|\frac{\Delta F(t)}{F_{avg}(t)}\right| = 2\left|\frac{F_1(t) - F_2(t)}{F_1(t) + F_2(t)}\right| \tag{10}$$

in which, $\Delta\sigma(t)$ and $\Delta F(t)$ are the differences of these two stresses and forces applied on the specimen end surfaces, respectively. $\sigma_{avg}(t)$ and $F_{avg}(t)$ are the averages of these two stresses and forces, respectively (see Eqs. (5) and (6)). The specimen is assumed to be in dynamic stress equilibrium, when the value of $R(t)$ is less than 0.05 [8]. This general criterion has been employed extensively to evaluate the dynamic stress equilibrium process in SHB experiments.

By substituting Eqs. (7) and (8) into Eq. (10), the parameter $R(t)$ can be expressed by means of the relation between the incident, reflected and transmitted waves, as below:

$$R(t) = 2\left|\frac{F_1(t) - F_2(t)}{F_1(t) + F_2(t)}\right| = 2\left|\frac{\varepsilon_I(t) + \varepsilon_R(t) - \varepsilon_T(t)}{\varepsilon_I(t) + \varepsilon_R(t) + \varepsilon_T(t)}\right| \tag{11}$$

For a specimen under dynamic stress equilibrium, the forces applied on the specimen end surfaces can be expressed in terms of the two pressure bar strains, as expressed in Eqs. (7) and (8). Thus, according to Eq. (11), the following relation can be derived:

$$\varepsilon_I(t) + \varepsilon_R(t) = \varepsilon_T(t) \tag{12}$$

Therefore, by applying an assumption of the positive travelling harmonic wave and the equation of motion, the expressions for the specimen's engineering stress, strain and strain rate can be derived in terms of the two pressure bars' strains, as shown in Eqs. (1), (4) and (2), respectively. In turn, under dynamic stress equilibrium, Eqs. (1), (4) and (2) are qualified to be used for calculating the specimen's engineering stress, strain and strain rate.

Therefore, the SHB technology for dynamic compression and tension experiments has been introduced. It is a mature method and a commonly used apparatus to characterize accurately the dynamic mechanical properties of various materials in the strain rates ranging from 1000/s to 10,000/s [20].

In this chapter, in order to characterize the high strain-rate mechanical response of polymer materials, the SHB technique is selected as the experimental equipment. On the one hand, the SHB equipment can accurately derive the dynamic mechanical properties of material under high-speed loading. On the other hand, it can be well controlled under laboratory working conditions [20].

3. Various polymers for dynamic mechanical investigation

3.1. Rubber-like polymers

Rubber-like polymer has been used extensively as structural material of engineering components that are designed to resist impact, ranging from bus windows and eyeglasses to protective helmets and body armours. The choice of polymer materials for these applications has been made appealing by the relative low density and the transparency that are the characteristics of amorphous homopolymers. One of the studies on polymer materials focuses on the capability of absorbing dynamic strain energy and strain-rate dependency [1, 21–30]. Strain-rate dependency of the stress-strain behaviour of polymer materials has been well documented, where, in particular, yield stress is found to increase with the increasing strain rate. This feature of mechanical behaviour is highly relevant to engineering applications, when designing a polymer component required to resist an impact loading.

3.1.1. Polyurea

The thermoplastic elastomer polyurethane and the elastomeric thermoset polyurea are found to have new applications by increasing the survivability of structures under impact loading, including those encountered in blast and ballistic events. Yi et al. [31] studied the large deformation and rate-dependent stress-strain behaviour of polyurea and polyurethanes in dynamic compressive tests. A set of data was presented to quantify the rate-dependent behaviour of these materials from low strain rates (<1/s) to high strain rates (>1000/s), as shown in **Figure 2**.

The polyurea displayed a transition of deformation behaviour from rubber-like behaviour at low strain rates to leathery behaviour at high strain rates, whereas one of these three polyurethanes displayed a transition from rubber-like behaviour at low strain rates to glass-like behaviour at high strain rates. **Figure 2** presents the rate-dependent behaviour by means of the stress vs logarithm strain rate, taking the stress evaluated at a strain level of 0.15 and 0.30 for

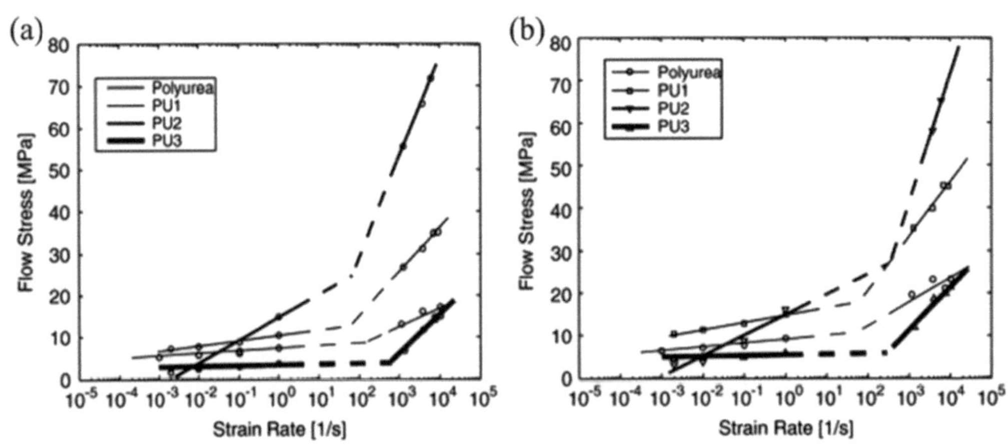

Figure 2. Flow stress vs strain rate relations of polyurea and three polyurethanes determined at the deformation strain of (a) 0.15 and (b) 0.30 [31].

http://dx.doi.org/10.5772/intechopen.69570

Figure 2a and **b**, respectively. For both strain levels, the flow stress obviously demonstrated a close-to-linear dependency on the logarithm strain rate in both the high strain rate ($\gtrsim 10^3$/s) and low strain rate ($\lesssim 10^0$/s) regions. So, mechanical behaviour, as illustrated by yield stress of the thermoplastic-elastomeric polyurethanes and elastomeric-thermoset polyureas, is strongly dependent on strain rate.

In addition, continuous investigation was conducted on the characterization of mechanical properties at very high strain rates under both dynamic compression and tension loadings [32]. The experimental results are shown in **Figure 3**. The uniaxial compression and tension data for polyurea are found to be consistent at the strain rates ranging from 0.001/s to 10,000/s, both of which increase with the increase in strain rate. Therefore, a strong dependency of flow stress on strain rate is clarified in the material of the thermoplastic-elastomeric polyurethane and elastomeric-thermoset polyurea, which is of particular interest when they play a role as a protective coating to enhance survivability of structures in high-rate loading events.

3.1.2. Polyurethane

Zhang et al. [33] studied the dynamic mechanical behaviour of a polyurethane used as an interlayer in a laminated windshield construction at various strain rates (0.001/s to 7000/s) and

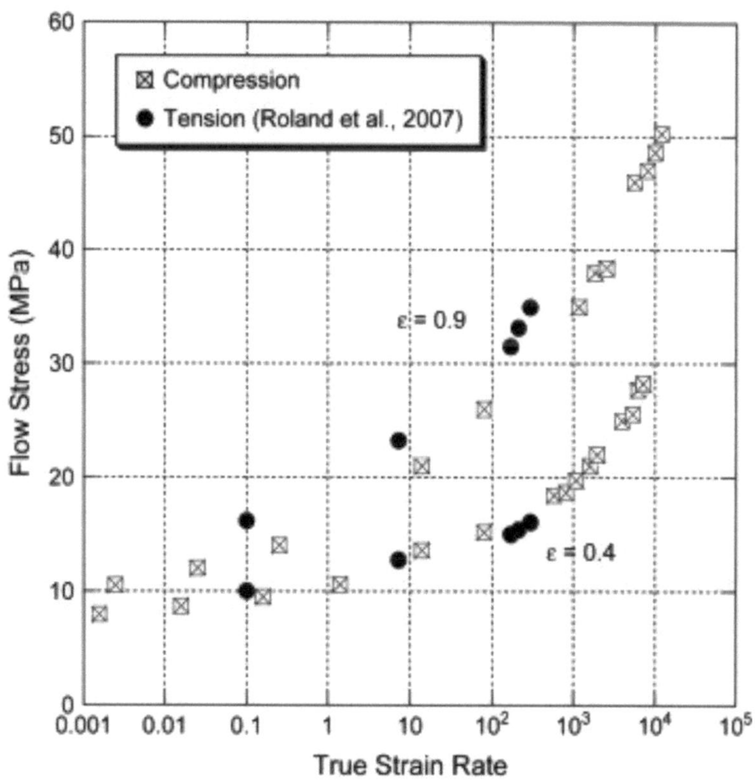

Figure 3. Stresses, taken at the true strain of 0.4 and 0.9, as a function of the true strain rate of the polyurea, where the value of the true strain rate used for each point is also taken at the true strain of 0.4 or 0.9 [32].

various temperatures (−40°C to 25°C). The research results reveal that the mechanical behaviour of polyurethane interlayer is dependent on temperature and strain rate. Under dynamic loading, the transition from "rubbery" to "glassy" is exhibited in the stress-strain curves at 240°C. In terms of the constitutive theory and experimental data, one-dimensional thermal-hyper-visco-elastic constitutive equation is recommended to characterize the compressive deformation response of polyurethane interlayer over a wide range of temperatures and strain rates.

Figure 4 shows the true stress-strain curves of polyurethane interlayer at different temperatures and at a certain strain rate. The temperature-dependent behaviour is clearly seen. When temperature decreases, the stress-strain curve goes up with the increase in yield stress, and the strain-hardening behaviour becomes remarkable. At high strain rates and at low temperature, the flow stress date exhibits an obvious increase. Under the quasi-static (0.001/s) loading, the stress-strain results illustrate a common phenomenological mechanical behaviour of soft materials revealed in compression experiment at a low strain rate. As the temperature decreases from −20°C to −40°C, the significant changes occur in stress-strain curves. A rubbery behaviour transits into

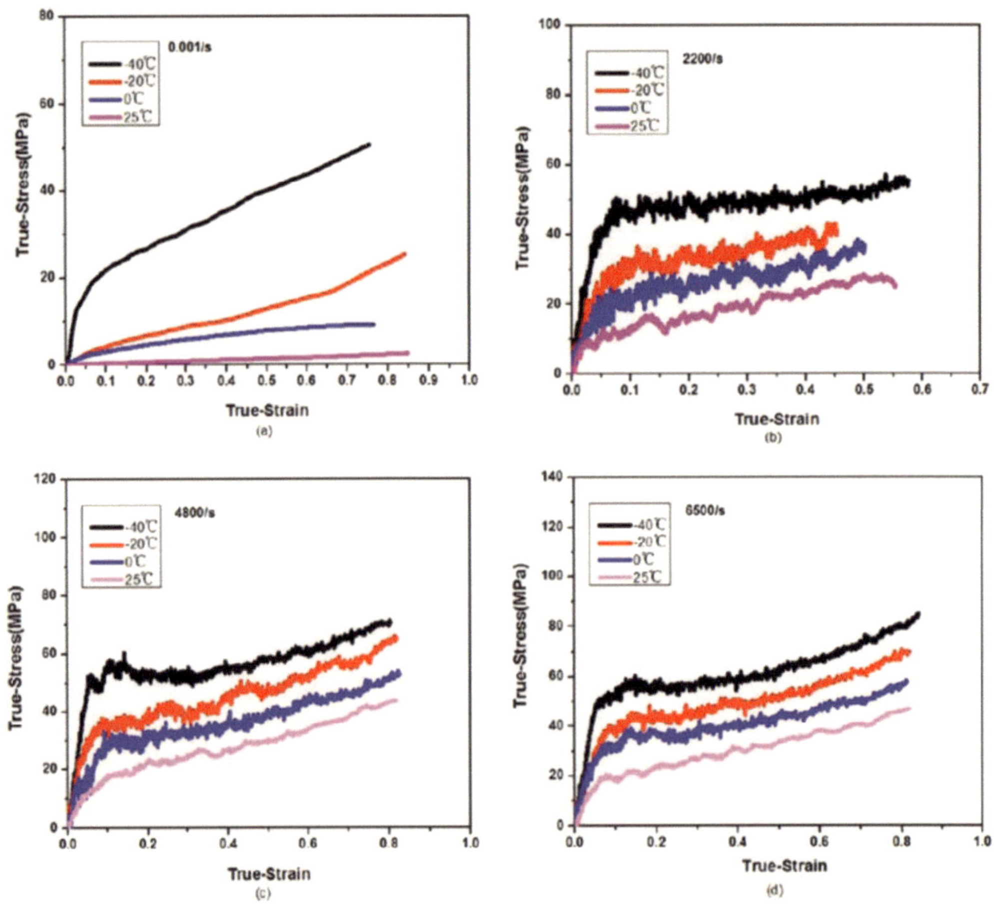

Figure 4. True stress-strain curves of the polyurethane interlayer: (a) 0.001/s, (b) 2200/s, (c) 4800/s and (d) 6500/s [33].

glassy behaviour of the mechanical response, which is in line with the mechanical behaviour of the rubber-like materials below and above glass transition temperature. The inherent reason can be the glass transition temperature of the polyurethane interlayer of around −58°C to −40°C.

Figure 5 shows the true stress-strain curves and the corresponding true strain rate-strain curves at a temperature of 25°C. The trend of strain rates seems to be relatively constant over the courses, which indicates that dynamic stress equilibrium is nearly stable. The uniaxial compressive stress-strain behaviour in the regime of high strain rate has a strong dependency of strain rate and temperature. At the same temperature, with the increase in strain rate, the flow stress increases and the strain hardening behaviour becomes more apparent.

3.1.3. Polyurethane elastomer

Fan et al. [34] developed a soft polyurethane elastomeric material for impact-resistant applications. Stress-strain relations, characterized by using a split Hopkinson tension bar, are derived to reveal the mechanical properties at different strain rate levels at room temperature, as shown in **Figure 6**. The stress-strain curves from multiple tests at comparable strain rates are similar and partially overlap, illustrating the good reproducibility of the experimental data. The dynamic stress-strain curves show a different behaviour, compared to quasi-static stress-strain plot at a strain rate of 0.01/s. This difference has been also observed for other soft polymer materials [35, 36].

In statics, the initial stiffness of the soft polyurethane elastomeric polymer material is much negligible. While in dynamics, stiffness becomes significantly higher, and the length of the linear ascending branch increases with the increase in strain rate. Even though stress equilibrium is not

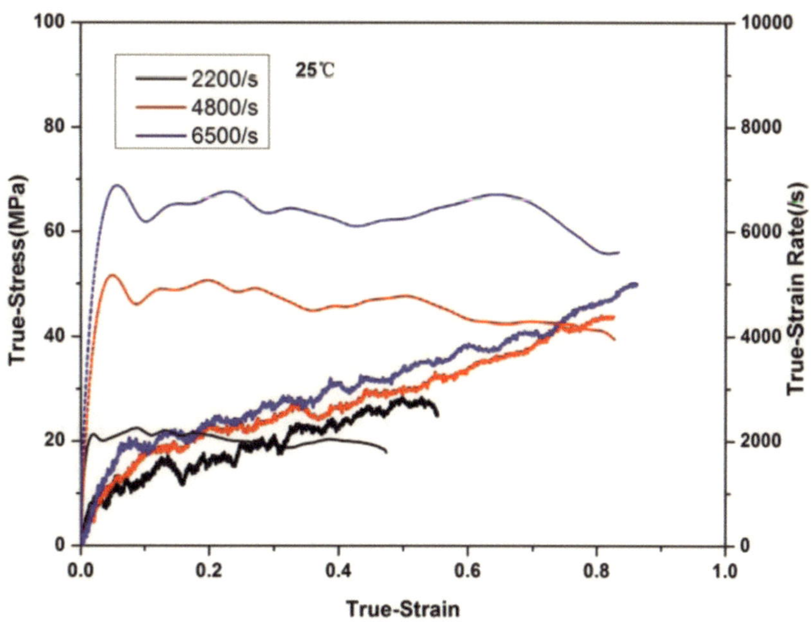

Figure 5. True stress-strain curves and the corresponding true strain rate-strain curves at a temperature of 25°C [33].

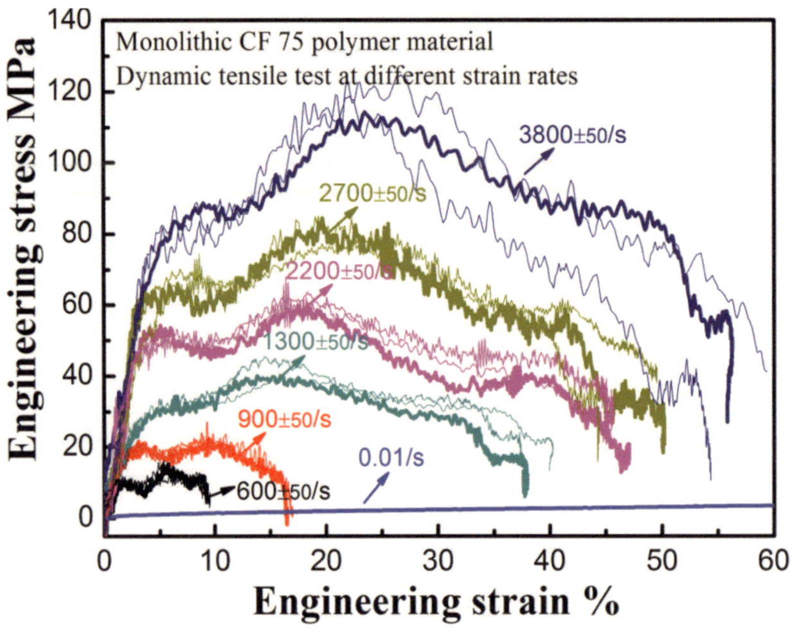

Figure 6. Representative engineering stress-strain plots of the soft polyurethane elastomeric polymer material under dynamic tension loading with three curves per selected strain rate level [34].

attained in the specimen at the beginning of the initial dynamic loading, a linear link of yielding point and origin point (0, 0) in the stress-strain curve can be conducted to roughly evaluate the dynamic tensile modulus, considering that material yielding occurs after dynamic stress equilibrium. The line slope is the tangent modulus, which can indicate the material stiffness at the corresponding strain rate. The relation between tangent modulus and strain rate is shown in **Figure 7**. By linearly fitting the curve of tangent modulus versus log strain rate, a log strain rate value of 2.65 or about 450/s strain rate is attained. It indicates that the strain rate of 450/s is the critical transition point at which mechanical response of the soft polyurethane elastomeric polymer material changes from a rubber-like behaviour at low strain rates to a glass-like behaviour at high strain rates at room temperature [31, 37].

3.2. Glass-like polymers

The development of glass-like polymer materials that are more impact- and ballistic-resistant has many possible applications ranging from military vehicle windows to civilian products. Two common kinds of organic glasses that are used in these engineering areas include the polycarbonate (PC) and polymethyl methacrylate (PMMA). These materials are transparent and lightweight compared with their inorganic counterparts, which is important especially in vehicle and personal protection applications. Each of these materials has a unique inherent mechanism by which energy is absorbed during impact. Polycarbonate, like other ductile materials, has the capability to absorb a large amount of energy through yielding and plasticity. For PMMA, the majority of energy absorption is due to the creation of new surface area during fracture. There is an obvious difference between their macroscopic failure mechanisms.

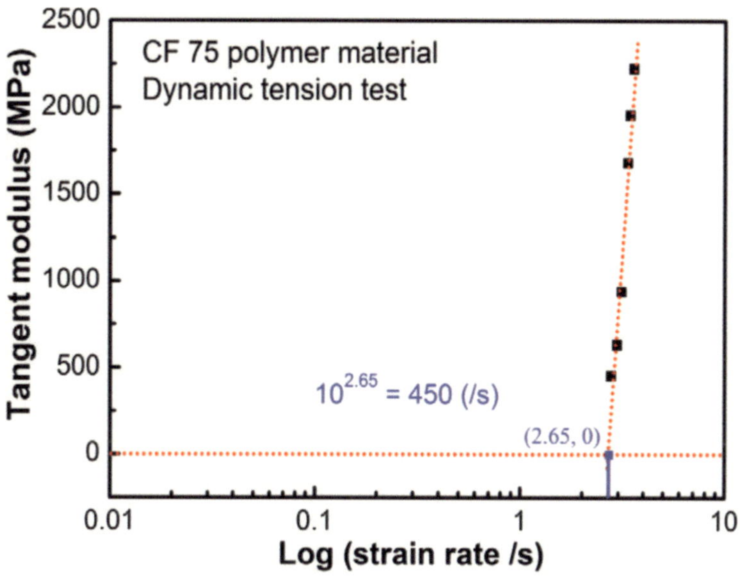

Figure 7. The relation of tangent modulus versus strain rate curve for determining the critical strain rate of the soft polyurethane elastomeric polymer material for the transition from a rubber-like behaviour at low strain rates to a glass-like behaviour at high strain rates at room temperature [34].

Failure in PC is relatively localized while PMMA is effective at delocalizing failure in the form of radial cracking and Hertzian cone fracture (during impact). Moreover, the glass-like polymers also have an obvious rate dependency of mechanical properties [1, 38, 39], which directly affects the dynamic strength and determines the structure design, application and reliability for safety and security.

3.2.1. Polymethyl methacrylate (PMMA)

As one of glass-like polymers, polymethyl methacrylate (PMMA) is reviewed for clarifying the dynamic mechanical behaviour [1]. A combined experimental and analytical method has been performed to investigate the mechanical behaviour of PMMA material at strain rates ranging from 10^{-4}/s to 10^{4}/s. The relation of yield stress and strain rate is documented in **Figure 8**.

The yield stress was found to increase as a non-linear function with the logarithmic strain rate, displaying the strain-rate sensitivity. The mechanisms of the rate-dependent elastic-plastic deformation of PMMA material from low to high strain rates were also studied. A computational model was developed based on the concepts of both the Ree-Eyring yield theory and the viscoelastic theory, which indicates that intermolecular resistance to deformation may be decomposed into the contributions of different molecular processes, each with their own unique rate and temperature dependency. This model is probably suitable for two-component polymer materials.

Therefore, rate-dependent behaviour of polymer materials was revealed and mechanisms for impact resistance were also explored using the combined experimental and computational

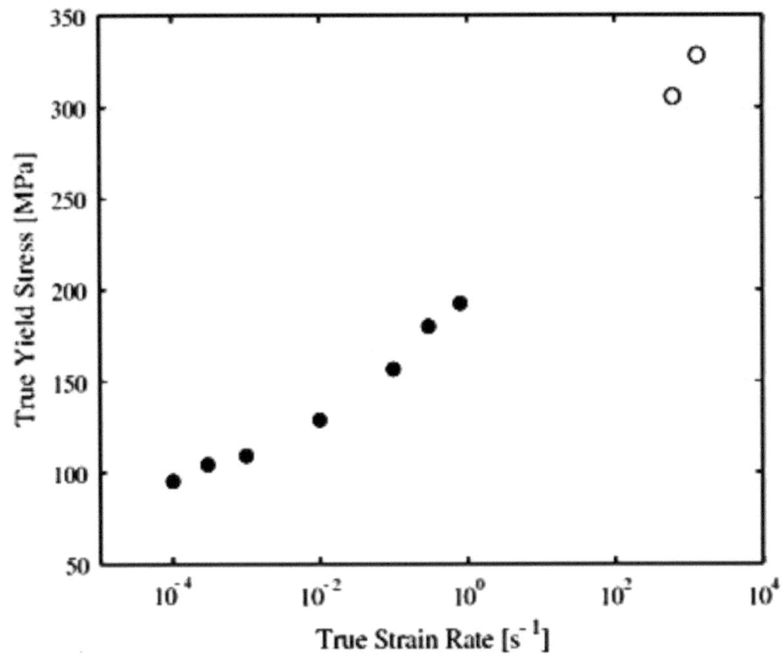

Figure 8. True yield stress of PMMA material as a function of true strain rate (logarithmic scale)—low to high strain rates [1].

methods. A concerted effort has been made to investigate and develop new polymer materials with improved characteristics for impact resistance and damage tolerance. Concerns regarding rate dependency and impact resistance of polymer materials are the basis for extensive research.

3.2.2. *Polycarbonate (PC)*

Dar et al. [39] studied the mechanical behaviour of polycarbonate (PC) polymer under the effect of various temperatures and strain rates. Mechanical characterizations are carried out through uniaxial compression and split Hopkinson pressure bar (SHPB) for revealing low and high strain rate response, respectively. Meanwhile, the experiments are performed for strain rates varying from 10^{-3}/s to 10^3/s and a temperature range of 213 K to 393 K. The experimental results reveal that the stress-strain behaviour of polycarbonates is much different at lower and higher strain rates. At higher strain rate, the polycarbonates yields at higher yield stress compared to that at low strain rate. At lower strain rate, yield stress increases with the increase in strain rate while it decreases significantly with the increase in temperature. Likewise, initial elastic modulus, yield and flow stress increase with the increase in strain rate, whereas decreases with the increase in temperature. Yield stress increases significantly for low temperature and higher strain rates.

SHPB tests were performed to determine the dynamic response of polycarbonate at strain rates varying from 1350/s to 9400/s. Dynamic tests were performed at five different strain rates, and the results in terms of true stress-strain curves are shown in **Figure 9**. The results show that yield stress increases with the increase in strain rate. The stress-strain curves show almost similar

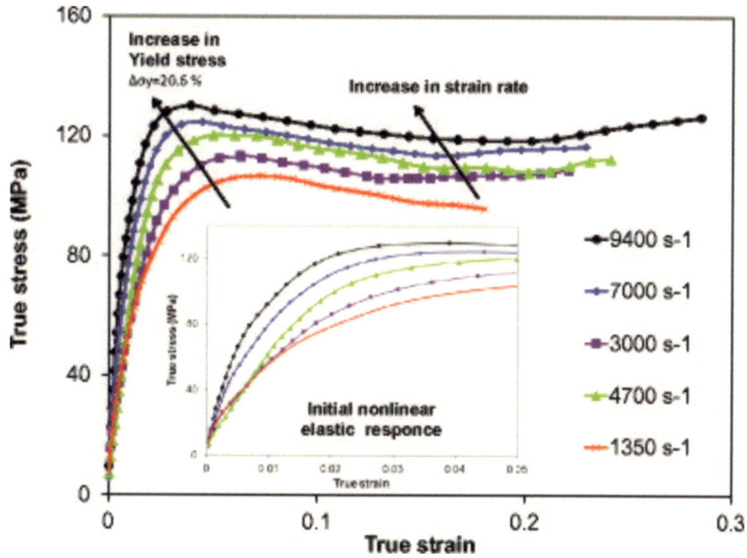

Figure 9. High strain rate stress-strain response of polycarbonate [39].

mechanical response in which initial nonlinear elastic behaviour was observed followed by subsequent yielding, strain softening and hardening. Yield stress changes significantly with the increase in strain rate. An increase of 20.6% in yield stress was calculated with strain rate increase from 1350 to 9400/s. At all strain rates, ductile response of polycarbonate was observed and ductile-brittle transition was not found.

Dynamic tests at a strain rate of 1350/s were also performed at three different temperatures and the results are shown in **Figure 10**. The change in yield stress is more significant in case of temperature than strain rate. The 43.4% decrease in yield stress with the increase in temperature from 233 K to 333 K is revealed.

Yield stress summarized at different strain rates and temperatures were plotted and a liner relationship was found between them as shown in **Figure 11**. A 0.69 MPa/K decrease in yield stress was observed between temperature variations of 233–333 K. Dynamic stress sensitivity $((\sigma_{dynamic} - \sigma_{static})/\sigma_{static}$ of polycarbonate is computed to be 128% which is significantly less than PMMA [40] showing that polycarbonate is not a highly strain rate-sensitive polymer. σ_{static} in this case is considered as quasi-static or very low strain rate stress.

3.2.3. *Polymethylene diisocyanate (PMDI)*

Song et al. [41] reported the dynamic mechanical response of three polymer foam materials made by rigid polymethylene diisocyanate (PMDI), varied in density (310 kg/m^3, 410 kg/m^3 and 550 kg/m^3), at strain rates as high as 3000/s and at temperatures ranging from 219 K to 347 K. The effects of material density, strain rate and temperature on the compressive response of the polymer foam materials were determined. Compressive stress-strain curves of the three foam materials (with the densities of 310 kg/m^3, 410 kg/m^3 and 550 kg/m^3) at various strain rates are shown in **Figure 12a–c**, respectively.

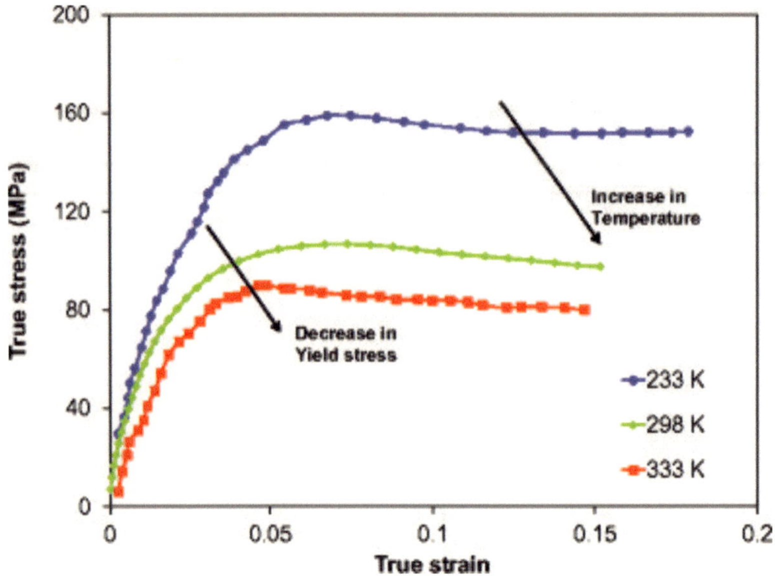

Figure 10. Effect of temperature on high strain rate stress-strain response [39].

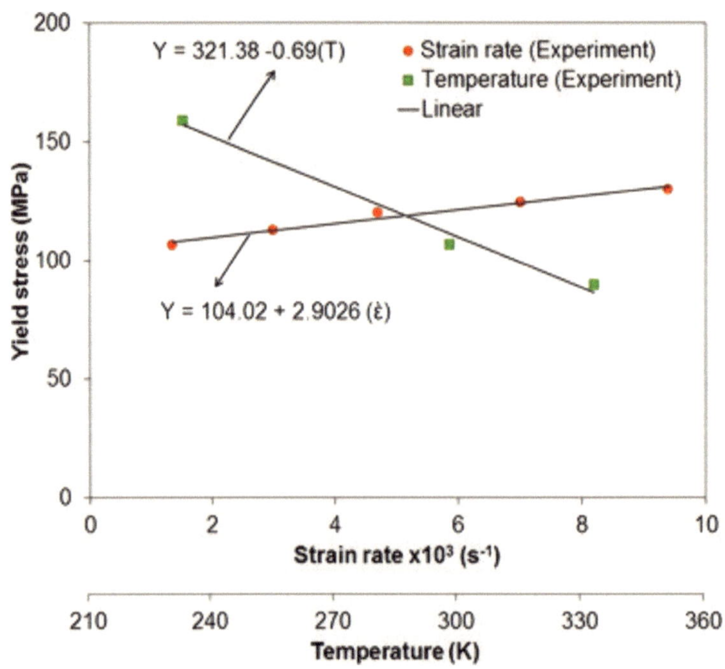

Figure 11. Effect of strain rate and temperature on yield stress [39].

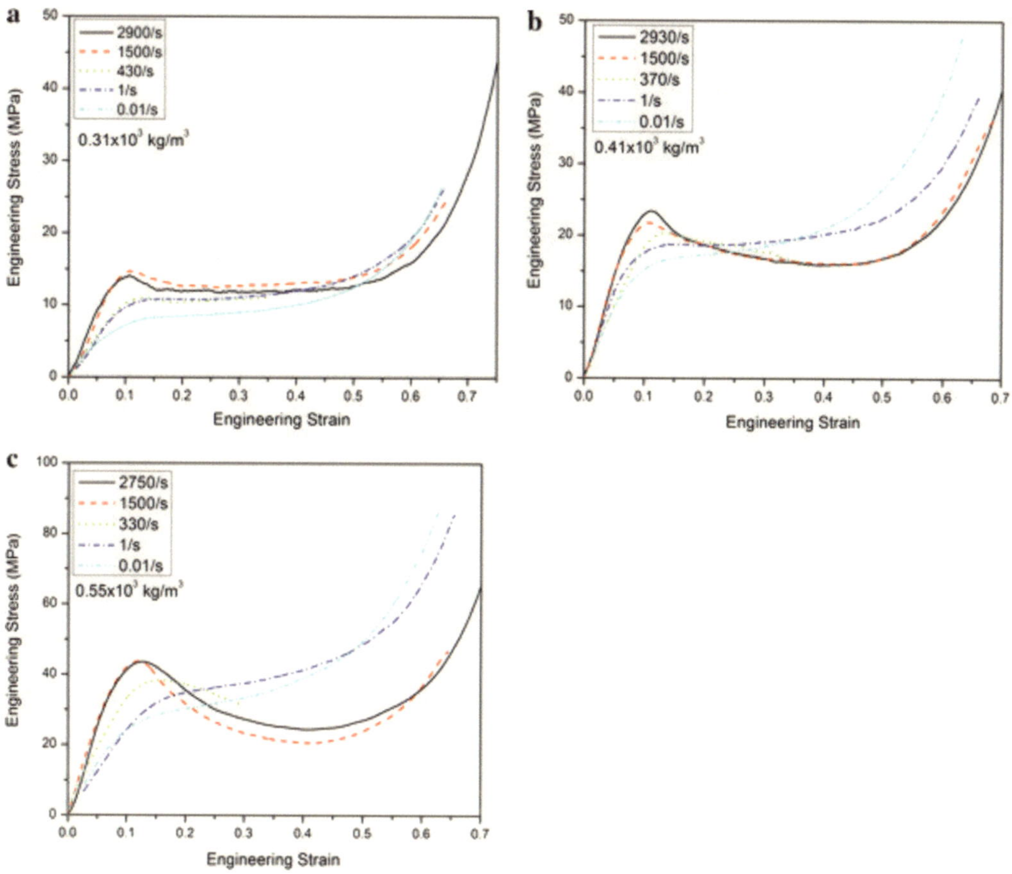

Figure 12. Compressive stress-strain curves at various strain rates for our three foam-material densities: (a) 310 kg/m^3, (b) 410 kg/m^3 and (c) 550 kg/m^3 [41].

They are obtained at room temperature (295 K) and have a common characteristic: an initial linearly elastic deformation followed by a collapse process of cell structures. When all the cell structures were collapsed, the condensation initiates, as revealed by the increasing stress amplitude in the stress-strain response [42]. For each polymer foam material, this characteristic is varied slightly with strain rates, and the cell-structure collapse process has an apparent variation. Under quasi-static loading, the stress-strain curves exhibit a long stress plateau and/or slow strain-hardening behaviour after yielding. This stress plateau indicates the plastic buckling of cell structures. However, under dynamic loading, stress drops from the peak date after yielding, causing the formation of N-shaped stress-strain curves. The stress drop is caused by the sudden collapse of cell structures under high strain-rate loading. So, the deformation and damage mechanisms are influenced by the strain rates in the polymer foam material.

Yield stress is also found to be dependent on strain rate. **Figure 13** shows the details of the increase in yield stress with strain rate for these three polymer foam materials. The yield strength of the three foam materials linearly increases with the logarithm of strain rate, as

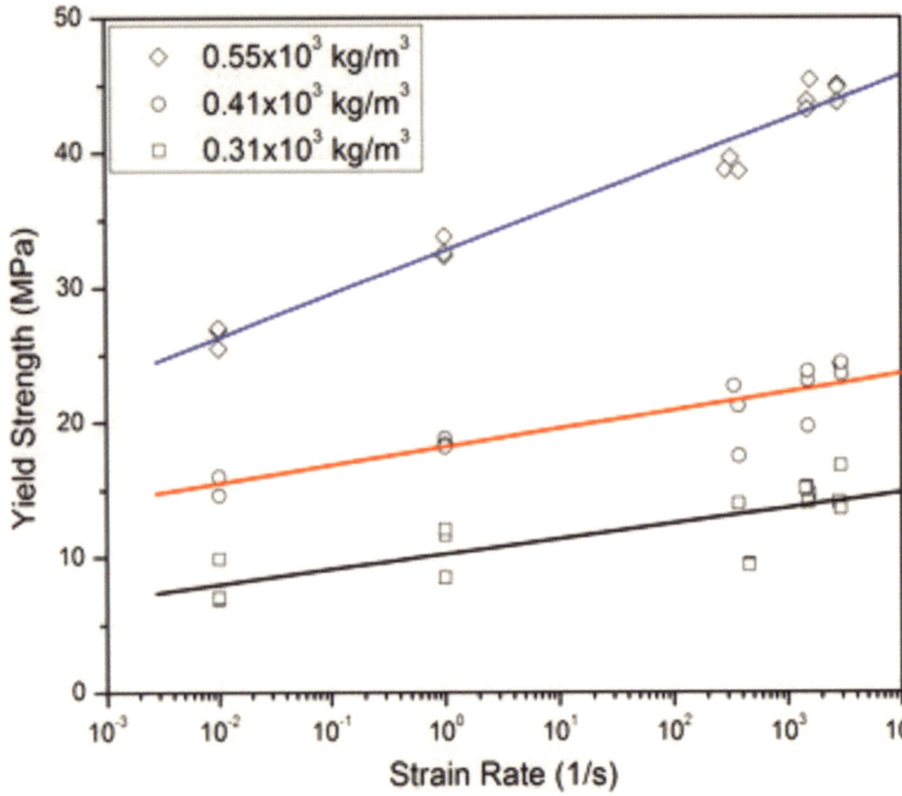

Figure 13. Strain-rate sensitivities of the three polymer foam materials with different densities [41].

shown by equation: $\sigma_y = A + Blog(\varepsilon'/\varepsilon'_0)$, wherein A and B are constant data, and ε'_0 is the reference strain rate. The constant B is the slope of the solid line, which represents the strain-rate sensitivity of yield stress. The two parallel lines at the bottom of **Figure 13** imply that the strain-rate sensitivities of the 310 kg/m^3 and 410 kg/m^3 polymer foam materials are nearly equal. Both are lower than that of the 550 kg/m^3 polymer foam material.

4. Conclusion

In this chapter, the characterization method of mechanical response of polymer materials under high strain-rate loading is firstly introduced. Then, two kinds of polymer materials, rubber-like and glass-like, are reviewed to illustrate their dynamic mechanical response. Herein, three polymers are presented as the representatives of each kind of polymer materials. The rate-dependent mechanical data are given and the influence of temperature is also clarified. These knowledge outputs not only guide the research of developing new impact-resistant polymer materials, but also support the protection engineering of applying polymer materials in dynamic events.

http://dx.doi.org/10.5772/intechopen.69570

Acknowledgements

The author, JTF, like to acknowledge the reuse permission of materials from publishers, and the financial supports of BIT-start-up funding, Project of State Key Laboratory of Explosion Science and Technology of QNKT17-03 and National Natural Science Foundation of China with Grant No. 11602024.

Author details

Jitang Fan[1,2]*, Xiaoyun Fan[1] and Ang Chen[1]

*Address all correspondence to: Jitang_fan@hotmail.com

1 State Key Laboratory of Explosion Science and Technology, Beijing Institute of Technology, Beijing, China

2 Advanced Research Institute for Multidisciplinary Science, Beijing Institute of Technology, Beijing, China

References

[1] Mulliken A, Boyce MC. Mechanics of the rate-dependent elastic-plastic deformation of glassy polymers from low to high strain rates. International Journal of Solids and Structures. 2006;**43**(5):1331–1356

[2] Richeton J, Ahzi S, Vecchio KS, Jiang FC, Adharapurapu RR. Influence of temperature and strain rate on the mechanical behaviour of three amorphous polymers: Characterization and modeling of the compressive yield stress. International Journal of Solids and Structures. 2006;**43**(7–8):2318–2335

[3] Roland CM, Twigg JN, Vu Y, Mott PH. High strain rate mechanical behaviour of polyuria. Polymer. 2007;**48**(2):574–578

[4] Naik NK, Pandya KS, Kavala VR, Zhang W, Koratkar NA. High-strain rate compressive behaviour of multi-walled carbon nanotube dispersed thermoset epoxy resin. Journal of Composite Materials. 2015;**49**(8):903–910

[5] Deshpande VS, Fleck NA. High strain rate compressive behaviour of aluminium alloy foams. International Journal of Impact Engineering. 2000;**24**(3):277–298

[6] Kolsky H. An investigation of the mechanical properties of materials at very high rates of loading. Proceedings of the Physical Society (London). 1949;**62**:676–700

[7] Grey GT III. Classic Split-Hopkinson pressure bar testing. In: ASM Handbook: Mechanical Testing and Evaluation. Metals Handbook. American Society for Metals, Materials Park. Vol. 8. 2000. pp. 462–476

[8] Chen W, Song B. Split Hopkinson (Kolsky) Bar: Design, Testing and Applications. Springer, New York Dordrecht Heidelberg London. 2011

[9] Tay TE, Ang HG, Shim VPW. An empirical strain rate-dependent constitutive relationship for glass-fibre reinforced epoxy and pure epoxy. Composite Structures. 1995;**33**:201–210

[10] Yang LM, Shim VPW, Lim CT. A visco-hyperelasitc approach to modeling the constitutive behaviour of rubber. International Journal of Impact Engineering. 2000;**24**: 545–560

[11] Gray GT, Blumenthal WR. Split Hopkinson pressure bar testing of soft materials. In: ASM Handbook: Mechanical Testing and Evaluation. Metals Handbook. American Society for Metals, Materials Park. Vol. 8. 2000. pp. 488–496

[12] Yin ZN, Wang TJ. Deformation of PC/ABS alloys at elevated temperatures and high strain rates. Materials Science and Engineering: A. 2008;**494**(1):304–313

[13] Bouix R, Viot P, Lataillade JL. Polypropylene foam behaviour under dynamic loadings: Strain rate, density and microstructure effects. International Journal of Impact Engineering. 2009;**36**(2):329–342

[14] Aleyaasiin M, Harrigan JJ. Wave dispersion and attenuation in viscoelastic polymeric bars: Analysing the effect of lateral inertia. International Journal of Mechanical Sciences. 2010;**52**:754–757

[15] Lee OS, Cho KS, Kim SH. Dynamic deformation behaviour of soft material using SHPB technique and pulse shaper. International Journal of Modern Physics. 2006;**20**(25–27): 3751–3756

[16] Rodríguez J, Cortés R, Martínez MA. Numerical study of the specimen size effect in the split Hopkinson pressure bar tests. Journal of Materials Science. 1995;**30**(18):4720–4725

[17] El-Habak AMA. Compressive resistance of unidirectional GFRP under high rate of loading. Journal of Composites Technology and Research. 1993;**15**(4):311–317

[18] Wang BZ, Hu SS, Zhou XR. Research of dynamic mechanical behaviour and constitutive model of rubber under different temperatures. Journal of Experimental Mechanics. 2007;**22**(1):1–6

[19] Ravichandran G, Subhash G. Critical appraisal of limiting strain rates for compression testing of ceramics in a split Hopkinson pressure bar. Journal of the American Ceramic Society. 1994;**77**(1):263–267

[20] Fan JT. Dynamic response of a polymer and polymer composite system: Experimental studies [PhD thesis]. Delft, The Netherlands: Delft University of Technology; 2015

[21] Balieu R, Lauro F, Bennani B, Delille R, Matsumoto T, Mottola E. A fully coupled elastoviscoplastic damage model at finite strains for mineral filled semi-crystalline polymer. International Journal of Plasticity. 2013;**51**:241–270

[22] Raman SN, Ngo T, Lu J, Mendis P. Experimental investigation on the tensile behaviour of polyurea at high strain rates. Materials & Design. 2013;**50**:124–129

[23] Plaseied A, Fatemi A. Strain rate and temperature effects on tensile properties and their representation in deformation modeling of vinyl ester polymer. International Journal of Polymeric Materials. 2008;**57**(5):463–479

[24] Daivis PJ, Matin ML, Todd BD. Nonlinear shear and elongational rheology of model polymer melts at low strain rates. Journal of Non-Newtonian Fluid Mechanics. 2007;**147** (1–2):35–44

[25] Yu Q, Selvadurai APS. Mechanics of a rate-dependent polymer network. Philosophical Magazine. 2007;**87**(24):3519–3530

[26] Hizoum K, Remond Y, Bahlouli N, Oshmyan V, Patlazhan S, Ahzi S. Nonlinear strain rate dependency and unloading behaviour of semi-crystalline polymers. Oil & Gas Science and Technology. 2006;**61**(6):743–749

[27] Perkins TT, Smith DE, Chu S. Single polymer dynamics in an elongational flow. Science. 1997;**276**(5321):2016–2021

[28] Smith DE, Chu S. Response of flexible polymers to a sudden elongational flow. Science. 1998;**281**(5381):1335–1340

[29] Mead DW, Larson RG, Doi M. A molecular theory for fast flows of entangled polymers. Macromolecules. 1998;**31**(22):7895–7914

[30] Arruda EM, Boyce MC, Jayachandran R. Effects of strain rate, temperature and thermomechanical coupling on the finite strain deformation of glassy polymers. Mechanics of Materials. 1995;**19**(2–3):193–212

[31] Yi J, Boyce MC, Lee GF, Balizer E. Large deformation rate-dependent stress-strain behaviour of polyuria and polyurethanes. Polymer. 2006;**47**(1):319–329

[32] Sarva SS, Deschanel S, Boyce MC, Chen WN. Stress-strain behaviour of a polyurea and a polyurethane from low to high strain rates. Polymer. 2006;**48**(8):2208–2213

[33] Zhang L, Yao X, Zang S, Gu Y. Temperature- and strain rate-dependent constitutive modeling of the large deformation behaviour of a transparent polyurethane interlayer. Polymer Engineering & Science. 2014;**55**(8):1864–1872

[34] Fan JT, Weerheijm J, Sluys LJ. High-strain-rate tensile mechanical response of a polyurethane elastomeric material. Polymer. 2015;**65**:72–80

[35] Yang JH, Zhang Y, Zhang YX. Brittle-ductile transition of PP/POE blends in both impact and high speed tensile tests. Polymer. 2003;**44**(17):5047–5052

[36] Song B, Chen WN, Liu ZS, Erhan SZ. Compressive properties of soybean oil-based polymers at quasi-static and dynamic strain rates. Journal of Applied Polymer Science. 2006;**99**(5):2759–2770

[37] Kendall MJ, Siviour CR. Rate dependence of poly(vinyl chloride), the effects of plasticizer and time-temperature superposition. Proceedings of the Royal Society A. 2014;**470**(2167): 20140012

[38] Acharya S, Mukhopadhyay AK. High strain rate compressive behaviour of PMMA. Polymer Bulletin. 2014;**71**(1):133–149

[39] Dar UA, Zhang WH, Xu YJ, et al. Thermal and strain rate sensitive compressive behaviour of polycarbonate polymer—experimental and constitutive analysis. Journal of Polymer Research. 2014;**21**(8):1–10

[40] Lee OS, Kim MS. Dynamic material property characterization by using split Hopkinson pressure bar (SHPB) technique. Nuclear Engineering and Design. 2003;**226**(2):119–125

[41] Song B, Lu WY, Chul JS, et al. The effects of strain rate, density, and temperature on the mechanical properties of polymethylene diisocyanate (PMDI)-based rigid polyurethane foams during compression. Journal of Materials Science. 2009;**44**(2):351–357

[42] Gibson LJ, Ashby MF. Cellular Solids, Structure and Properties. 2nd ed. Cambridge: Cambridge University Press; 1999